P9-AFB-743

URBAN

ABRAMS, NEW YORK

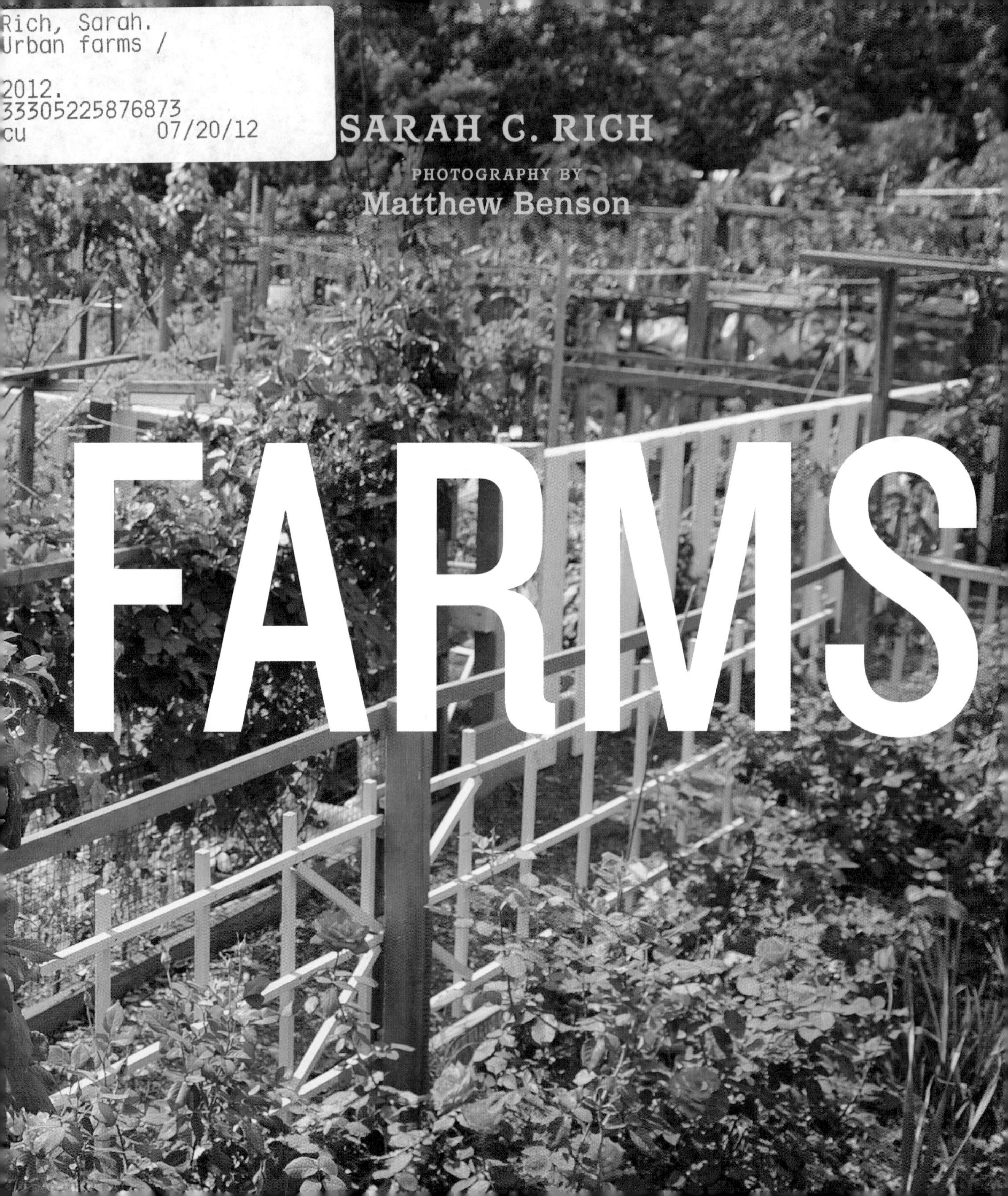
SARAH C. RICH
PHOTOGRAPHY BY
Matthew Benson
FARMS

CONTENTS

12 INTRODUCTION

18 VICTORY PROGRAMS ReVISION URBAN FARM

30 BROOKLYN GRANGE

42 EDIBLE SCHOOLYARD NYC

54 Edible Education
by Allison Arieff

58 GREENSGROW FARM

74 Edible Entrepreneurship
by Makalé Faber Cullen

78 COMMON GOOD CITY FARM

92 FARNSWORTH

102 EARTHWORKS URBAN FARM

110 Growing Public Health
by Rupal Sanghvi

114 CATHERINE FERGUSON ACADEMY

120 CHICAGO CITY FARM

132 GROWING POWER

152 The Art of Growing Food
by Nicola Twilley

156 OUR SCHOOL AT BLAIR GROCERY

170 HOLLYGROVE MARKET AND FARM

178 Little Homestead in the City
by Alissa Walker

182 NEW ORLEANS HOMESTEAD

192 GHOST TOWN FARM

204 WATTLES COMMUNITY GARDEN

212 MILAGRO ALLEGRO COMMUNITY GARDEN

220 BIOGRAPHIES

222 ACKNOWLEDGMENTS

BROOKLYN GRANGE

INTRODUCTION

It was a fortuitous miscalculation that led this book's photographer, Matthew Benson, and me to drive through Chicago one autumn day and pay a spontaneous visit to the crew at City Farm. We were traveling from Will Allen's legendary Growing Power farm in Milwaukee to the urban agriculture mecca that is Detroit—a road trip far longer than we'd bargained for—and as we circled Lake Michigan, we agreed we'd be remiss in our exploration if we didn't make a stop in a city long recognized as a sustainability leader. With the indispensable assistance of a cell-phone-to-laptop wireless connection, I sat in the passenger seat as we sped south on I-94 and tracked down City Farm manager Andy Rozendaal online. Within a few minutes we were making plans to meet.

The visit to City Farm was inspiring for many reasons—a profusion of vegetables burst forth from neatly designed garden rows, which lie at the base of a strikingly modern single-resident-occupancy building, designed by architect Helmut Jahn, and against a backdrop of the downtown Chicago skyline. Residents of the SRO, many of whom are recovering from substance addiction, take a share of the farm's output and even help out from time to time. The day we visited, a trio of kittens roamed the tomato vines, in training to run pest control. On one corner of the property, bees hovered lazily around a stacked hive. We took many great photos and learned all about the history of City Farm, but perhaps the greatest takeaway of the day was Rozendaal's response to a question I was asking throughout our travels: What is the difference between an urban garden and an urban farm?

If you wake one morning to dismal weather, Rozendaal told me, and you can stay indoors sipping something warm, watching the sleet or hail pound the soil, and wait for fairer weather before stepping outside, then what you have is a garden. But if your livelihood depends on your crop and you roll out of bed each day and head straight out to your land, rain or shine, then you are the steward of a farm. According to Rozendaal, it is not the size of the plot or the presence of animals, but the need for sustenance—nutritional and financial—that makes a farm a farm.

While not every urban farm included in this book represents the sole source of subsistence for the people working there, all of these projects share the philosophy that agriculture can and should sustain a city, whether that means feeding and strengthening struggling communities, catalyzing economic growth and creating job opportunities, improving public health and environmental conditions, or some combination of these.

Like many cities that are now sprouting gardens in place of asphalt, Chicago's modern farmers can trace a rich legacy of practitioners and visionaries who saw agriculture as part of an ideal metropolis. English social theorist and political activist Ebenezer Howard lived in Chicago in the 1870s, two decades before writing one of the seminal works of urban-design thinking, *To-Morrow: A Peaceful Path to Real Reform*. The book, which was later printed under the title *Garden Cities of To-morrow*, laid out an integrated city plan that knit swaths of productive farmland and open space into the populated and industrial fabric of the then-modern city.

In 1956, architect Frank Lloyd Wright proposed a building for Chicago that would have been the world's tallest skyscraper, designed to house the population of a small city on a vertical axis, leaving the horizontal plane available for farms. The ground-level portion of his vision was more akin to today's suburbs than to the cities we now know, and the engineering hurdles required to execute his plan rendered it nothing more than a futuristic fantasy. But it was a fantasy that foretold our present moment, when the availability of advanced technology, the growth of urban populations, and concerns over food security and environmental decline are all directing our attention to the promise of introducing agriculture within the city limits.

Urban farming is a uniquely powerful tool for change, in that it can simultaneously reshape the places where we live and the way we eat. It is also uniquely accessible—available to grassroots change agents and high-ranking policymakers alike. Just since the beginning of 2010, the launch of Michelle Obama's "Let's Move" campaign and her White House

GREENSGROW, PHILADELPHIA

VICTORY PROGRAMS REVISION URBAN FARM, BOSTON

garden, the presentation of the TED prize to chef and food activist Jamie Oliver, and the release of the FoodNYC report for urban agriculture in New York City all signaled serious intentions to change Americans' relationship with food, starting with how and where it's produced.

For Obama, Oliver, and the many other leaders and citizens who are now joining the movement, a key step in this process involves getting fresh food into urban neighborhoods where currently none is available. In these so-called food deserts, it is often easier to plant vegetables than to alter the inventory of the mini-marts and liquor stores where many people purchase food or to entice bigger supermarkets to risk opening in low-income neighborhoods.

Agricultural interventions not only facilitate access to healthy food, they become a new social anchor for the community. In New Orleans's Lower Ninth Ward, instead of loitering around the convenience store or playing in the street after school, neighborhood youth gravitate toward Our School at Blair Grocery, a small farm on a formerly neglected corner lot, which has been transformed into a safe haven by an educator named Nat Turner, who relocated from Brooklyn after Hurricane Katrina. Ninth Ward kids can now spend their afternoons feeding chickens, watering sprouts, or simply doing their homework amid the greenery. Blair Grocery farmers even built a small aquaculture structure for raising fish—a rich source of protein and an integral part of any traditional New Orleans menu.

In Detroit, Earthworks Urban Farm extends the services of the Capuchin Soup Kitchen, inviting resource-strapped citizens to grow some of their own food and infusing the soup kitchen's daily offerings with fresh, local produce. In a city suffering from economic recession and population decline, Earthworks also acts as a green jobs training ground, preparing people to earn a living as farmers, food distributors, and market employees.

Where job infrastructure does not already exist, agriculture becomes an engine for entrepreneurship. One of the great start-up stories to emerge from Queens, New York, in the last few years begins on a rooftop, where an ambitious group of local food advocates laid down a one-acre farm with a primary goal of establishing a fiscally sustainable business. Brooklyn Grange farm manager Ben Flanner earns a full-time salary digging in the dirt within a stone's throw of Manhattan. The farm's revenue stems primarily from selling produce at local farmers' markets, directly to restaurants, and through a community-supported agriculture (CSA) program.

The CSA model has proven to be an effective way to generate regular, year-round income and a useful method for connecting urban farms with suburban and rural operations. In Philadelphia, Greensgrow Farm runs a thriving CSA, with members of all backgrounds coming from around the city to pick up their weekly box of local food. In order to offer a diverse variety of vegetables, fruits, and value-added products, Greensgrow partners with farms outside the city, enabling them to include dairy, baked items, and even locally raised meat on their roster of items.

If the difference between a garden and a farm really does boil down to living off the land, then perhaps the quintessential characters in the urban farming cast are modern homesteaders—individual city dwellers who are attempting to grow and make everything they need within the boundaries of their own property. Between the creative spirit of the DIY movement and the apocalyptic tenor of news about climate change and economic collapse, there seems to be a perfect storm motivating people to experiment with self-sufficient living.

Pioneering homesteader Novella Carpenter has built a tiny empire of urban agriculture in the shadow of an Oakland freeway overpass. Around her dilapidated duplex she houses goats, rabbits, ducks, chickens, and even raised and slaughtered a three-hundred-pound hog. Her accounts of surviving off whatever her city lot would yield paint a picture of extreme dedication, occasional sacrifice, and real empowerment.

The urban farming leaders depicted in this book are just a representative sampling of a movement that is thriving from coast to coast. The sense of autonomy that comes from growing food and raising animals is a welcome experience at a time when the forces of the world often feel beyond individual control. Whether motivated by health concerns, environmental issues, financial constraints, or simply the creative challenge of a DIY endeavor, farming the city is an affordable, accessible way to contribute to a healthy urban future—and to eat well along the way.

VICTORY PROGRAMS ReVISION URBAN FARM

BOSTON, MASSACHUSETTS

In the history of urban farming in America, Boston occupies a significant place as the home of Fenway Victory Gardens, the only continuously operating original garden remaining of the Victory Garden movement that flourished during World War II. Dating back to 1942, this legacy of self-sustenance has inspired Boston residents to keep backyard gardens and encouraged the municipal government to support local food initiatives.

Just as mid-twentieth-century Victory Gardens demonstrated the power of growing food to strengthen community, some of Boston's more contemporary agricultural efforts have been instrumental in showing just how many facets of society can benefit from planting vegetables and tending bees. One thriving example emerged from the social services nonprofit Victory Programs, an organization founded in 1975 to provide treatment and housing for Boston's homeless, substance-addicted, or chronically ill citizens.

Victory Programs ReVision Urban Farm began at ReVision Family Home, a shelter for twenty-two families. In 1990, some of the mothers at the shelter decided to plant a small vegetable garden to grow nutritious food for themselves and their kids. The garden quickly became a vehicle for education and empowerment for the residents of the home, and their enthusiasm spurred Victory Programs to purchase a cluster of vacant lots across the street from the home. Using soil from a local cemetery and composted manure from a local zoo and the Boston Police Department's fleet of horses, residents and community members primed the land for optimal sun exposure, drainage, and aesthetic integration into the neighborhood. "The farm became a vibrant tract of land surrounding the shelter and helped create beauty where there was blight in the neighborhood," says farm manager Jolie Olivetti.

Today, Victory Programs ReVision Urban Farm cultivates a full acre with the helping hands of three full-time staff and a rotating crew of shelter residents, as well as local volunteers and youth. "Three main goals have shaped the development of our ReVision Urban Farm," Olivetti says, "small-scale, green, economic development; community food security; and job training and education. Community-based food production is a means for individuals to regain control over what we eat and how we grow it."

Several of the ReVision Family Home residents have gone on to join the farm staff or to create their own vegetable gardens wherever they live

Victory Programs
revision
urban

after they've left the shelter. Many of the youth who train with the farmers pursue environmental education, urban agriculture, and food justice in college and beyond.

Much of the food they produce goes to the residents at the shelter and to pantries around the city, but increasingly the farm also generates some revenue through sales of their produce, as well as of vegetable and flower seedlings, which they sell to home gardeners. In 2010, Victory Programs also opened a public farm stand to make fresh food available in the neighborhood, which has long been a food desert. With support from the mayor, they are able to offer steeply discounted prices to customers purchasing with food stamps or other government assistance.

During the summer, Victory Programs partners with Allandale Farm, a larger operation located in the Boston suburbs, to offer a twenty-week community-supported agriculture membership. Their signature tomatoes, collard greens, and squash are supplemented by a wide variety of crops grown on Allandale's acreage. During peak season, CSA members come from all over the city—from affluent and low-income neighborhoods alike—to partake in the collaboratively harvested bounty.

"Many aspiring gardeners and farmers visit us, volunteer with us, and learn from us," says Olivetti. "And we rely on the network of urban and peri-urban farmers to recommend best practices on everything and anything, from what to do about aphids to who's the best produce packaging supplier." A few years ago, she recalls, an elderly woman living near the farm leaned out of her window to critique Olivetti's sweet potato harvesting technique. She was welcomed into the garden to demonstrate her preferred way to pull the tubers and, before long, "we pulled these enormous, two- and three-pound beauties from the earth, laughing and shouting with the effort of digging out our treasure."

Start a Neighborhood Yield-Sharing Network

Sharing food with neighbors is an age-old way to build community, but the Internet and social networks have introduced a whole new approach to locating and organizing people interested in sharing the food that grows in their yards. From Craigslist, Facebook, and Twitter to smaller networks such as NeighborGoods, we are now lucky enough to have an infrastructure in place for spreading the word about grassroots efforts. If you are interested in gleaning fruit from neighborhood trees or redistributing excess vegetables during the height of summer, a combination of online and offline channels can get the word out quickly. Pick a central location and convenient time, and invite your neighbors to bring their yield to a swap. Those without gardens or fruit trees can enjoy a taste of fresh, local produce, and those with too much on their hands have an easy, free way to prevent waste.

SOUTH
BEAC

FARM FRESH
Vegetables
Vegetables
FARM FRESH
Vegetables

BROOKLYN GRANGE

QUEENS, NEW YORK

If you are looking for the rooftop farm known as Brooklyn Grange, you'll have to look in Queens. Atop a largely vacant, seven-story building along Northern Boulevard in Long Island City, rows of vegetables span a full acre that was, until spring 2010, just like every other industrial rooftop in the borough: an expanse of unexploited potential. To access the roof, you must traverse cavernous, empty rooms brightly lit during the day by wide warehouse windows. A metal staircase leads up to the rooftop door that spits you out among okra stalks and tomato vines.

Brooklyn Grange was founded by five young Brooklynites pursuing sustainability in both agriculture and business. Cofounder Ben Flanner, who works full-time as the head of the operation, got his start at Eagle Street Rooftop Farm, another rooftop venture, before striking out in search of a larger piece of real estate. The search itself was an education in urban property dealings and building codes. "As an endeavor like this is in completely uncharted territory," Flanner explains, "there was no precedent set for the value of a roof, length of a lease, and so on. We had lots of interested parties, but it was difficult to close the deal."

They launched the project before finalizing an agreement on a building, ultimately landing in a different borough from the one for which the business is named. But it's just a hop from Brooklyn, and it's from there, where local farming and small-scale food production are so ardently embraced, that many of the volunteers and visitors come.

Flanner himself has a degree in industrial engineering and worked in consulting and finance for five years in New York City before deciding to become a farmer, which he views as "an interesting way to apply constant problem solving and analytical skills." Indeed, a day at Brooklyn Grange is full of both, from managing the even distribution of drip irrigation to calculating the right cost scales for the farm's CSA.

The year before starting the farm, Flanner and his cofounders kept meticulous notes on yields and sales at the urban garden located at the popular Brooklyn restaurant Roberta's, owned by another member of the founding team, Chris Parachini. "I kept a large Excel file that allowed us to model out, for example, what the farm would look like at six times the size," he says, "or if we increased the growing area or slightly tweaked the vegetables to grow more profitable crops."

WESTERN QUEENS
COMPOST
INITIATIVE

Eventually they were able to calculate profit and loss for a year of farming a half-acre rooftop and to develop an operation plan that would allow them to pay Flanner a modest salary and still break even. While the space they ultimately rented was twice that of the one they'd modeled, the plan was solid enough to be scalable. They raised $200,000 in capital from investors, loans, and even through the online crowd-funding site, Kickstarter. Then they signed a ten-year lease and got to work.

One of the advantages of rooftop farming in New York, as opposed to, say, California, is that rooftops are engineered to withstand the significant weight of heavy snow, which makes them more likely to be farm-ready at the outset. An engineer's report on the Brooklyn Grange roof determined that no additional reinforcement would be needed to accommodate the garden beds. Still, the material that they chose to haul up is lighter than standard soil—a composite of porous stones and compost intended especially for green roofs. The soil mixture sits atop a protective base layer that collects water and shields the building structure.

With full exposure to the elements and plenty of heat in the summer, the rooftop is a great place for growing tomatoes, eggplant, peppers, and other sun-loving vegetables. They also do a big business in leafy greens of all kinds and keep bees at the edge of the site for pollination and honey. The Brooklyn Grange business model relies on sales to restaurants, their CSA membership, and market sales. In the lobby of their Queens building, they operate a small market of their own, in addition to selling produce at other markets around New York.

When the weather is good, Brooklyn Grange also runs regular events in the evenings, bringing the community together around big, family-style meals served on long tables among the rows of mustard and arugula. Guests pay a set price and local chefs prepare several courses using produce from the roof and artisanal products from around the city. The Manhattan skyline forms a glittering backdrop as the sun goes down.

It's not hard to find skeptics in New York and beyond who doubt that an urban farm—let alone any small farm—can be profitable, but the team at Brooklyn Grange has an exceptionally strong commitment to making a business case for everything they do, and so far it has paid off enough to keep skeptics at bay. In their very first year, the farm turned a small profit, allowing them to keep the promises they'd made to lenders and investors.

Their dedication to financial sustainability means that they do not casually allow visitors up to witness farming in action. While many community farms are open to the public and fairly casual about giving tours, the farmers at Brooklyn Grange found that this policy detracted from their productivity, and ultimately their profitability, and they decided to charge visitors to access the rooftop. Unfortunately, this meant that their aspiration to present an educational opportunity to local students would be difficult to realize.

As a solution, the ambitious team decided to add a nonprofit arm to their for-profit enterprise, which would be dedicated to making educational programming viable. City Growers aims to get kids into the dirt at Brooklyn Grange so they can learn about growing, harvesting, and eating local food within their own city limits.

While farming is an occupation that requires solid focus on immediate, day-to-day demands, the founders of Brooklyn Grange also take a long view on their future. Once the first site has proven itself to be profitable, they envision replicating their model on other New York rooftops, building a network of farms that can produce healthy food for the five boroughs. "I think New York City makes a lot of sense for this type of project because it has such a large and dense population, good weather, and plenty of large, strong roofs," says Flanner. "Also, New York has emerged as a location where people are interested in eating wiser and doing their small part to take care of the food chain."

STONE TILES
Northern Boulevard is
KOEPPEL
"Where Customer
Satisfaction is
YOUR
AD!
"WE KNOW THE NEIGHBORHOO
EXIT

Start Your Own Vegetable Garden

When you're ready to try growing your own food, there are several ways to start. You can purchase young plants from a nursery or a local garden, or you can begin from scratch with seeds. There's something to be said for the satisfaction of participating in the complete growing cycle—it's both inspiring and educational to watch a sprout emerge from the soil and become something delicious to eat. Seed catalogs, such as Johnny's Selected Seeds, Baker Creek Heirloom Seeds, or Seeds of Change, tend to present the greatest array of varieties and now offer online purchasing as well, or you can buy packets at a garden store or even collect seeds from a favorite heirloom vegetable that you want to grow yourself. Starting your seeds is easy to do on a kitchen windowsill or in a greenhouse. The incubation area should be between 65 and 70 degrees—to keep it consistent, an electric bulb can be brought to bear.

Begin with a plastic or biodegradable container with drainage holes at the bottom—clean yogurt cups or egg crates work well. Fill the container with a soil-free potting mix or a combination of compost, peat moss, and perlite. Wet the "soil" and compress it gently before inserting seeds. Seeds should be placed shallowly, at a depth of two or three times the height of the seed. Keep them covered with plastic to maintain the heat and moisture of the soil, and avoid exposing them to bright, direct sunlight. However, once they sprout, sun exposure is important, and as soon as there's no risk of frost outdoors, the small plants can be taken outside to acclimate and eventually be transplanted directly into the ground. A few common starter plants for the home gardener include herbs like basil and oregano, lettuces and other leafy greens like kale, and of course, the peerless homegrown tomato.

Northern Boulevard is
KOEPPEL
"Where Customer
Satisfaction is
YOUR
AD!
STONE TILES
NEIGHBORHOOD

Public Storage
Public Storage
Carrier
Carrier

16
16
17
16

EDIBLE SCHOOLYARD NYC

BROOKLYN, NEW YORK

The southern end of Brooklyn is probably best known for being the site of Coney Island, the famous beach resort and amusement park that had its heyday in the late nineteenth and early twentieth centuries. Besides its significance as a great historic leisure destination, Coney Island represents a major turning point for New York City, when manufacturing and mass transit began to dominate a previously agrarian landscape. In fact, the construction of passenger train lines in the late 1800s enabled Coney Island's success. While the economy prospered through industrial growth, the borough's farming culture faded away.

But the regional roots haven't been forgotten, and in the quiet residential neighborhoods of South Brooklyn, vegetable gardens are common, particularly among immigrants who bring agricultural practices from their native countries. While many of these residents are getting on in years, a new public school project in the Gravesend neighborhood is ensuring that the youngsters growing up here develop knowledge and enthusiasm for cultivating the land.

Edible Schoolyard NYC, located at P.S. 216, the Arturo Toscanini School, is the first East Coast affiliate of the celebrated program founded in Berkeley, California, by pioneering chef and food activist Alice Waters. The original Edible Schoolyard has become an important model for integrating farming and food education into a standard school curriculum. While each national affiliate is distinct according to regional climate and geography, the educational template for Edible Schoolyard NYC has been greatly informed by the fifteen years of experimentation at the West Coast site.

"It's critical that we reinforce existing curricular core standards of New York interdisciplinary education," says Christiane Baker, the program's executive director. "The Edible Schoolyard is a venue for every kind of traditional learning—reading, language arts, math, history—not to mention community building and social development. You can learn anything through the lens of a garden."

P.S. 216 serves almost five hundred children, from prekindergarten through fifth grade. The student body is ethnically diverse, and most kids come from low- and moderate-income families who qualify for food assistance at a rate well above the citywide average of 41 percent—according to the department of education, almost 70 percent receive public assistance. Edible Schoolyard NYC aims to close some of these disparities as a positive

consequence of teaching kids and their families about growing and eating nutritious, affordable food.

Farm manager Vera Fabian leads four classes per day throughout the year, ensuring that each student spends two consecutive days per month in the garden. "We bring them out two days in a row so that there's sustained learning and a balance of academics with hands-on garden work," Fabian explains. On the first day, the kids might spend the hour studying geography as a prelude to understanding seasonal eating, while on the second day they get their hands dirty in a composting workshop or learning how to harvest lettuce. When they return to their indoor classrooms, the garden pops up again in the process of learning to make graphs and charts tracking vegetable growth.

From the Edible Schoolyard perspective, lunchtime is a class too, or at least a great opportunity to extend edible education to its logical conclusion on the plate. P.S. 216 participates with the New York City Department of Education's SchoolFood program, which brings culinary school graduates to public school cafeterias to work with staff in preparing fresh, healthy meals from scratch. Through this program, the cafeteria is able to offer students access to a salad bar, whole grains, and, when possible, farm-to-table meals prepared using garden produce harvested on-site. The Garden to School Cafe also works with P.S. 216 in creating lunch menus that correspond with the local seasons for fruits and vegetables.

Baker and Fabian note that one of the most significant benefits they've observed among the students is an increased sense of empowerment and self-confidence. For approximately half of the student body, English is a second language—the P.S. 216 demographic includes kids from Central Asia, Eastern Europe, the Middle East, and China. The classroom can be a daunting environment for students with a limited ability to express themselves, but outside in the garden, learning to identify a flower or describe the flavor of a carrot can provide a more organic language acquisition opportunity. In a few instances, reports Fabian, children who do not have a firm grasp of English turn out to have a secret mastery in the garden, learned before immigrating or from grandparents. Growing food gives them an opportunity to shine and to teach their peers.

The garden plot is relatively small—just a quarter-acre of soil converted from an asphalt parking lot. Fabian designed the layout in sections conducive to teaching and based on listening to the desires of the students and their parents. "The kids requested grass so that they could be barefoot," she says. "So we're making a little space for that." An adjacent hoop house provides a sheltered environment for continuing garden education during colder months.

In 2012, a permanent greenhouse and kitchen classroom will fill out the rest of the schoolyard. Designed by New York–based architecture firm WORKac, the greenhouse and kitchen compose a self-sustaining facility that generates its own energy and collects rainwater that can be used to irrigate the gardens. Eventually, the sound of clucking chickens will mix with the din of children's voices.

Once per month, the school also offers Family Nights, where students come with their parents to learn how to cook the foods they're growing and studying in the garden and kitchen classroom. Shared meals provide a community-building opportunity for busy parents who may not have regular chances to meet the families of their children's peers.

The P.S. 216 project is the first of what will ultimately be a network of Edible Schoolyards in New York City. Baker aspires to create a citywide mandate for edible education, with showcase farms in each borough and professional development programs to equip teachers and administrators to establish their own farms and gardens. Perhaps the best indication of the city's support for this effort came from Brooklyn borough president Marty Markowitz, who told Baker that Edible Schoolyard NYC was the first public project he'd been involved with about which nobody had a complaint. "Gardening is the great equalizer," Baker says. "Everyone is an eater, and nothing builds community like growing food together."

JEOPARDY!
PLANT PARTS

EDIBLE EDUCATION

BY ALLISON ARIEFF

ALL IMAGES IN THIS ESSAY FROM EDIBLE SCHOOLYARD NYC (PAGE 42)

Remember home ec? And its male corollary—shop class? Girls learned to keep a clean house, prepare a meal, and balance a checkbook, while guys got to make birdhouses and use circular saws. The former was decried—understandably for the era—by feminists as the enemy of gender progress; the latter was scuttled as boys ditched hand tools in favor of digital screens and joysticks. The shift away from vocational skills toward more test-based learning was initially viewed as a positive step toward getting kids on the college track, but it has become clear that we've lost something rather integral to the process of cultural, social, and intellectual development: basic skills.

The culling of classroom minutes devoted to simple tasks like changing oil or sewing a button may have played a part in successfully designing a curriculum as a means to an end: study subject, pass test, go to college. But it had the unintended consequence of rendering most Gen X and Yers helpless in the quotidian realm—reliant on their moms to do laundry and on the microwave to cook their meals. With all eyes on prepping for Harvard, even the tying of shoes has become a lost art as nearly every preschoolers' laces have by now been replaced by Velcro tabs.

We can compensate for much of lost basic life skills, but the act of doing, of making, of watching raw materials merge into a thing or a meal—that's been in short supply in classrooms. The introduction in the last decade of school garden programs has been a pragmatic move in the right direction, familiarizing youth with the origins and growth processes of their food while instilling in them a willingness to get dirty and cultivate a meal with their own two hands.

While learning to garden seems innocuous enough, not everyone has been in favor of adding it back into the school day. In the *Atlantic* in 2010, Caitlin Flanagan argued that sending kids out to till the soil represents a step back for education:

> "On the first day of sixth grade, the boy walks through the imposing double doors of his new school, stows his backpack, and then heads out to the field, where he stoops under a hot sun and begins to pick

lettuce. . . . A cruel trick has been pulled on this benighted child by an agglomeration of foodies and educational reformers who are propelled by a vacuous if well-meaning ideology that is responsible for robbing an increasing number of American schoolchildren of hours they might otherwise have spent reading important books or learning higher math. . . ."

I remember reading this essay and thinking that clearly the author had never been present when a kid had dug up and admired "ugly" potatoes or seen him marvel at the wriggling of fat earthworms that emerged when he had done so. As one online commenter to her essay put it, "Are you truly positing that having grade school students spend one to two hours a week outside growing vegetables and fruit plants in a garden is a huge drain on the California educational system?"

Certainly not every child will take a shine to gardening, just as not every student loves reading or math, but school garden programs have proven themselves an invaluable educational resource. Like the home economics and shop classes of yore, they teach something practical to be sure, and they also celebrate collaboration, community, and the simple satisfaction that comes from following something from start to finish.

Any good teacher knows that children learn not just from books and tests, worksheets and quizzes, but through projects, interaction, and experiences. Though it's been decades since I sat in her book- and travel-ephemera-filled classroom, I still remember with remarkable clarity my sixth-grade teacher, Mrs. Bergeron. Among the year's worth of engaging class projects, she had us build our visions for a City of the Future (mine was a George Jetson–worthy landscape of upended plastic champagne glasses spray-painted silvery blue). Today, I write about architecture, design, and urbanism for a living. Coincidence? Perhaps, but no worksheet or pop quiz would have led me here.

In urban areas in particular, children are spending more and more time inside and less time being active. More screen time and less exercise are perhaps inevitable trends as technology-oriented learning increases and public funding for physical education drops, but some of the ripple effects are worth attention and prevention.

Consider the distinctly twenty-first-century problem of what Richard Louv has described as nature-deficit disorder, wherein children spend less and less time outdoors, leading to a wide range of behavioral problems. The past decade has seen an alarming spike in childhood obesity, juvenile diabetes, antisocial behavior, attention deficit and/or hyperactivity disorders, and there's little sign of a change to this trend. Can a school garden program eliminate all these ills? No, but it sure can help.

First and foremost, school garden programs get kids outside and begin to reacquaint them with the natural world. Once there, they're getting dirty (another increasingly rarified activity in our antibacterial-obsessed world) and discovering where food comes from and how it grows. Artificially lit supermarket year-round abundance yields to an utterly different reality as kids discover what grows when (and what it looks like before it's processed and packaged). They become attuned to weather patterns, seasons, and the behavior of bugs, good and bad. From there, it's not a huge leap

to talk about everything from phases of the moon to entomology to nutrition and healthy eating.

Anything that grows can become the focus of a classroom curriculum as proven by REAL School Gardens in Texas which developed its very own Potato Scholars Program to engage students in math, science, language arts, nutrition, and environmental education. Once harvested, the kids' potato crop is donated to their local food bank. As an elementary school teacher who participates in one of REAL School Gardens' seventy-four programs explains, the garden "is central because we do reading, writing, and arithmetic out there. Every day we're collecting scientific data . . . they learn about volume, ratio, rates of application. . . . We look at a rain gauge and talk about [measurement] and the fractions of an inch. . . . We then start breaking that into decimals. It's a very integrated curriculum with math and science."

Utilized with care and creativity, school gardens provide real opportunities to combat troubling trends regarding children's learning, health, and parental support. "School gardens are more than pretty places on campus . . . [they're] living laboratories for investigations into real-world problems and provide practical opportunities to grow a new generation of innovators," executive director Jeanne McCarty adds. There's also strong evidence that experiential learning programs like REAL Schools make learning fun without sacrificing educational objectives and academic performance. Further, at a time when consumption of fruits and vegetables is at a treacherous low for many youth, research has shown that children who grow their own food are more likely to eat it. Nearly 17 percent of the students of the REAL Schools are obese, and empowering them to make better food choices on their own is one of the best ways to facilitate healthy habits that will stick around through adulthood and get passed down when those students have their own children. That self-motivated decision making is key, and it stands to reason that when people start making better choices for their own health, they are also more inclined toward thinking about the health of their community and the environment. A six-year-old is more likely to wear the shirt she picked out herself; she's also more inclined to eat the carrot she picked from the ground herself.

Developing school garden programs is very much about creating interactive learning spaces. As Rebecca Lemos of Washington DC's City Blossoms explains, "[These spaces are] for children to use their creativity and combined strength and skills to learn how to grow and maintain fantastic yet functioning gardens."

These gardens, she continues, provide opportunities to learn about time management and the process of planning out a project step-by-step, as well as teaching careful observation, problem solving, and handling the challenges of managing a living system, where sometimes plants fail to thrive and produce. Unlike instant test results, gardening offers children situations in which they can think creatively in the long term; it is a project that progresses over a continuum and in the context of natural forces and seasonal rhythms.

A program like City Blossoms emphasizes the relationship between schools and their respective communities. There are other options as well, such as Partner Gardens, which works with the children, staff, and members of a school or community organization to create and maintain a garden, and also Community Green Spaces, a program whereby large plots of land are loaned to community members by private owners.

Lemos and her colleagues have been amazed by the kids' increasing awareness of the many elements of the natural world around them, the diversity of the different foods and plants that can be grown, the interconnectedness of their world and the world outside. "In our student gardeners," she says, "we see growing desires to access and use green spaces in their communities, which they see as theirs to shape and share."

The effects are not necessarily instantaneous, cautions Lemos, but, she explains, "We have noticed that gardens are usually the inspiration to create more changes in their environments, whether it is at a school or community center, that will influence attitudes of the community as a whole—for example, starting recycling in the classrooms, choosing healthy snacks that can be made using ingredients from the gardens, and wanting to spend more time outside interacting with nature or with each other."

In Los Angeles, the United States' second largest city, the Garden School Foundation operates a three-quarter-acre vegetable and quarter-acre native garden in the schoolyard at 24th Street Elementary School and runs classes for the students five days a week. Each of the nearly one thousand students attends six seed-to-table cooking and nutrition classes every year in which they plant, harvest, and prepare fresh produce from the garden, finishing the class by sitting down to eat together. Interviews done after the close of the six classes, says GSF's Julia Cotts, "show an increased participation in what students cook at home, including more cooking with their families, as well as increased demand for produce from the supermarket."

As is the case with most successful school garden programs, the Garden School Foundation is tailored to the needs of the community it serves. As Cotts explains, "For children living in poor urban areas in Los Angeles, nature is extremely scarce. Those open green spaces that do exist are very often plagued by crime and aren't safe for children to enjoy. First and foremost, gardens are a way for children to connect to a nature that has disappeared from our city. Additionally, schools are increasingly drilling students solely to achieve on tests while extracurricular activities are defunded. Gardens give children hands-on, experiential learning opportunities where they get to investigate, explore, learn for themselves, and teach each other, building essential life skills they don't have the opportunity to learn anywhere else." Addressing issues of food access is key: Obesity and diet-related disease are a huge epidemic in these areas, says Cotts, and "gardens are a surefire way to get kids to eat more vegetables. There's nothing more true than the mantra 'If they grow it, they'll eat it,' something we see during every one of our cooking and nutrition classes conducted in the garden."

School gardens can't end urban poverty or class inequality, restore strapped education budgets, or eliminate childhood obesity, but they're having an impact. They're changing attitudes and habits, fostering collaboration and new ideas, and creating community. As the quixotic artist and writer Maira Kalman observed on her *New York Times* blog when she visited the kids at Berkeley's Edible Schoolyard, "They pick beans and kale and pineapple guavas. They roast peppers. They churn butter. And they cook. And then they sit down together and eat and talk. And philosophize. . . . Then they fold the tablecloth. And sweep. And do all the things that families have been doing for hundreds of years."

GREENSGROW FARM

PHILADELPHIA, PENNSYLVANIA

When Mary Seton Corboy started Greensgrow Farm in Philadelphia in 1998 with Tom Sereduk, urban farming—and sustainability in general—were not yet household words. And the corner of the city where she chose to set down roots had not previously been a bastion of greenery. The traditionally working-class neighborhood of North Philadelphia had seen the rise and fall of various manufacturing industries, and much of the land was contaminated, including the three-quarters of an acre on which Greensgrow was born.

"The EPA had cleaned the lot up and put up a fence and gate around it and left," Corboy recounts. "It sat as a short-term dump site after that for about five years—nothing here but rubbish." The community around the farm was mostly white and working-class, with many people unemployed and, according to Corboy, having a not uncommon distaste for outsiders. "They thought we were insane," she says of their arrival, and not everyone was welcoming. "But they've gotten used to us."

Greensgrow operates as a for-profit business, and making a living off the land is important to Corboy, who has spent more than a decade building the farm to the point where that's possible. A DC native, Corboy didn't have an agricultural background when she started, so Greensgrow provided an immersive environment for learning the ropes of gardening. "Common sense and frugality were our guides," she says.

When Greensgrow began, they focused only on growing lettuce, which continues to be one of their primary outputs, though they have expanded to include other vegetables. All Greensgrow produce is planted in raised, hydroponic beds, which require no soil and have a permanently recirculating water system that keeps resource use to a minimum. Overhead remains modest as well, since the farm sends almost no wastewater back into the municipal system.

In recent years, Greensgrow has installed living roofs atop the permanent structures on the farm site, to further prevent water runoff. Planted with hardy sedum and wildflowers, the green roofs attract beneficial insects to the farm and support the bee population that lives in the Greensgrow apiary. In one of these permanent structures, the Greensgrow nursery runs year-round, providing a spot for ambitious Philadelphia residents to source vegetable starters for their own home gardens, purchase plants, or pick up cut flowers.

Welcome to
greensgrow!

Today, Greensgrow employs about twenty people, most of them recently out of college. In her dry, humorous way, Corboy expresses doubt that a high percentage of them will still be digging in the dirt when they're her age. The relentless labor and relatively low pay can lose its romance after a time. A few years of farming weeds out the dabblers from the lifelong farmers. "I don't believe all these kids who say they want to be farmers," she says. "They'll want something else when they are thirty. Urban agriculture is part of the solution but a darn small one."

But her attitude perhaps underestimates the difference her own small farm has made in Philadelphia and the cumulative impact of the many small farms that have sprung up during the years since Greensgrow began, galvanizing a clear movement toward restoring local food production in cities.

At the Greensgrow site, a lively farm stand draws visitors from around the city who are eager to get their hands on the freshly picked lettuces and herbs grown there, and other goods procured from nearby, including artisanal products like cheese, bread, and pasture-raised meat. Greensgrow also creates their own value-added products, which they make in a local church kitchen and sell throughout the year. Their goods include jams, vegetable spreads, sauces, and seasonal pies—all of which enable them to use produce that is past its prime for raw consumption but still plenty valuable when cooked or preserved.

Another major component of the Greensgrow business is their CSA, which in their case stands for "city-supported agriculture." The year-round program provides members with a box full of produce, dairy, and even occasionally a local specialty item like a Pennsylvania craft beer. Further bolstering the local food ecosystem, Greensgrow works with Philadelphia restaurants to collect waste grease from commercial kitchens, which they convert to biodiesel to run the trucks they drive to collect their inventory from around the region. It's a comprehensive model that grows denser each time Greensgrow partners with another local company or producer. While the farm itself keeps its focus narrow, cultivating a small variety of vegetables and herbs, they are the hub of a network that showcases the many efforts in and around Philadelphia to create sustainable small-scale farms and food businesses.

But for all of Corboy's own dedication as an entrepreneur, her outlook for the future of food at a larger scale is not particularly optimistic. "Business is a dirty word," she says, acknowledging the common friction between the idealism of small-scale farming and the realities of achieving success in a capitalist economy. "It's too bad," she continues, "because it's the only way to survive." She believes that farm subsidies are likely inevitable, despite their negative impact on small farms, and wears no rose-colored glasses about the possibility that small farms will feed a starving and rapidly growing global population. Of course, many would disagree with her bleak prediction for the planet's future, but few people can criticize the work she's doing in her own corner of the globe, where an urban Philadelphia neighborhood has become several degrees healthier and more livable thanks to the presence of her farm.

"People like Greensgrow. We've grown into our own work boots and never overcapitalized, so we owe no money, have no debt, and do as we please," she says, reflecting on the growth of the farm over more than a decade. "And being slightly older than most urban farms means more work, less talk. To paraphrase Kipling: Farms are not made by sitting in the shade saying, 'Oh, so lovely.'"

← 2400 - 2500 →
FIRTH STREET
← 2500 - 2599 →
GAUL STREET
Farm
Stand

CUMBERLAND ST
ONE WAY

PLANT MATTER HERE
DONT TOUCH
COMPOST GUIDELINES!
*ONLY PUT MATERIAL THAT WILL DECOMPOSE QUICKLY INTO COMPOST BINS (no branches, stems, vines, or roots.)
*SPREAD NEW MATERIAL OUT, THEN COVER WITH COMPOST, SAWDUST, OR STRAW
*PUT ANY X-TRA MATERIAL/MATERIAL
WILL NOT EASILY
PLANT MATTER HERE

Composting
Toilet
Farm Stand
Perennials
LIVING ROOF

CSA
D'Ambrosio
KEEP REFRIGERATED
Brand

Welcome to
Greensgrow Farm
Sale

BEECHWOOD

STOP

6
Fig Tree
1
Solar Panels

EDIBLE ENTRE-PRENEURSHIP

BY MAKALÉ FABER CULLEN

ALL IMAGES FROM GREENSGROW FARM (PAGE 58)

In 2002 John William Kulm, a Washington State farmer, published a book of poems and essays entitled *The Five Stages of Quitting Farming*. In it, he shared that farming is sublime. That it's absurd. That it's a calling. John's tormented good-bye to what would've been a fifth generation family legacy mirrored thousands of other practical and unfair good-byes as America's foundation industries of farming, logging, ranching, and mining were shuttered, spitting out their worn and tired laborers as they closed. John looked to the nearest city to try his hand at a more predictable job.

Around the same time that rural family farms like John's were going into the steep decline of the 1980s, people in cities started growing their own food again. They did so in response to the parallel demise of America's urban manufacturing, distribution, and food processing industries and the blight they left behind. A collective called Green Guerillas seed-bombed vacant lots with sweet peas, tomatoes, and butterfly weed. Their mission was to clean, green, and beautify city neighborhoods and to relearn basic life skills. Their motivations were political, economic, and environmental.

Out of these subversive activities have come new enterprises. The Green Guerillas and others like them have been propelled forward with knowledge culled from diverse sources, such as tips from immigrants who planted kitchen gardens on fire escapes and created backyard orchards, sometimes skirting outdated legal statutes to create something meaningful and useful. City zoning be damned or catch up with the times. As U.S. trade policy moved agricultural production and jobs to "the global periphery," seeking cheap labor and new consumers overseas, small groups of intrepid U.S. residents began taking it upon themselves to re-anchor food production and jobs at home.

Do urban farms yield jobs? Yes. A highly diverse one-acre neighborhood plot can generate income for one or more individuals. But it's revealing to talk about jobs in a broad sense. Urban farms also provide volunteers with skills-building opportunities that often help their careers in other sectors, and they enhance the value, safety, and beauty of a neighborhood as well, which, in turn, builds residents' self-esteem and earning potential.

There is also an interest in farm-work outside of blighted urban communities. Educated residents living in well-equipped neighborhoods are working the fields as well. Some are on their way to a career on a rural farm, others to becoming proprietors of an artisanal grocery, and increasingly, some have an urban farm as their ultimate career destination. For everyone it seems to be about "Satisfying the Hunger for Meaningful Work"—the tagline of Good Food Jobs, a national food and farm job placement service cofounded in 2010 by a farmer's daughter from Kentucky, Dorothy Neagle, and a New Jersey native, Taylor Cocalis (aka the Gastrognomes). Their pace stays breakneck to keep up with demand. This is in contrast to the U.S. Bureau of Labor Statistics' 2008–2018 occupational projections, which lead us to believe that only the construction and service-providing industries are on the rise in America. Agricultural work

Apples
Honey Crisp
Greensgrow
Sale
40 % off
Tropicals and
Pequea Valley Farm
Blueberry Yogurt
Chocolate Yogurt
Lemon Yogurt
Cherry Yogurt

prospects, they say, are static at best but generally on the decline because of high imports and the ability of the U.S. agriculture industry to achieve higher outputs with less labor. But we don't eat output. We eat food.

One of the many challenges urban farmers face is the need to develop relevant production protocols. "We borrow from authentic farming practices, but how do we sustainably deal with our city pests?" asks Linda Bryant, the feisty artist and director of the Active Citizen Project (ACP), which is incubating seven community farms throughout New York City. "We have to make peace with the rats! How do we peacefully coexist with our wildlife? And other human life?"

The veteran team running Greensgrow Farm in North Philadelphia since 1998 seems to prefer these types of challenges. For one, they say good land is readily available and affordable in cities when leased. Urban farmland is also closer to the city farmers' markets, which allows farmworkers to easily direct-sell and cuts transportation costs that whittle away profit. And, "it's easier to find farm labor in urban areas than in rural areas. There are a lot of very smart, motivated, and hardworking people who want to work in urban agriculture. The biggest problem with urban labor for agriculture is the lack of training. There just aren't training programs yet for urban agriculture—but it will come as the industry grows. We're all still just trying to figure out how it works," writes Mary Seton Corboy, Greensgrow's chief farmhand. (For more on Greensgrow, see page 58.) The wait won't be long. Urban farm production protocols have been developed internationally and are being exchanged throughout North America through sites like cityfarmer.org and Grow Youngstown urban farm in Ohio, which regularly provides an affordable daylong training—Hands-On Urban Farming: Sustainability and Profits in Small Spaces—in partnership with the National Sustainable Agriculture Information Service and others.

There are many practical and idealistic motivations for pursuing urban farmwork. But one typically prevails: Life in America is expensive. Housing, health care, education, fuel—there's little relief from the inflation. And then there are food prices. The International Monetary Fund determined that between March 2007 and March 2008, global food prices rose 43 percent. "I paid attention during the early part of the food crisis [of 2009]," reflects ACP's Bryant, who started an initiative called EATS at the Active Citizen Project, which facilitates food access and healthy eating through art and new media. "I saw Haitians eating cookies made from dried mud and a drizzle of honey. I thought to myself, money aside, what would happen to the eight million people in New York City if we lost access to, say, the bridges that brought food to us? Could we feed ourselves?" That, after all, is our basic human job.

Bryant's EATS urban farms do two things: produce food and distribute food. In doing so they're also creating a new workforce with the skills to feed itself. For three years an entire community is taken on for a farm apprenticeship. Workers receive training at their local farm lot—at Gorzynski Ornery Farm in Sullivan County, New York, for instance—and they design and run their farm stand retail operation. Several teenage EATS participants finishing high school at the Brownsville, Brooklyn, EATS farm have said that they thought they'd go into the military or nursing, but now they want to see if they can go to college for agricultural science.

Bryant is pleased, but only partially satisfied. "The most important project and the biggest challenge for us as urban farmers is to cultivate a shift in people's consumptive food behavior so people who live where food is grown are actually eating it." In observing, surveying, and listening to community members, EATS found that most participants consume prepared foods. And so, EATS partnered with Kingsborough Community College, home to CUNY's Center for Economic and Workforce Development, a commercial kitchen incubator and a culinary degree program. Now EATS community markets have a product line that includes not only just-picked vegetables but also ready-to-cook and ready-to-eat foods like beet applesauce made by culinary students at Kingsborough who are also community members with a cultural taste sensitivity. The products are even packed by young food manufacturing entrepreneurs.

High-quality ingredients used to make good-tasting prepared foods have actually propelled a food crafting renaissance in cities around the United States, which can be a pathway to urban farming. Take Brooklyn's Michael Cosaboom:

> "My decision to start a cottage industry coincided with buying a home with my wife and the birth of my youngest kids. I felt like our mortgage was so

big that the house needed to pay for itself somehow. I had never had unlimited roof access in a home before and I had done some previous experiments with container gardening and drip irrigation, so I began to experiment on my own roof. My vague notion was that I wanted to start a business that my kids can participate in someday. Over a few growing seasons I focused on the chiltepin family of peppers and started growing in quantities that I had a surplus of. I formed an LLC, developed a recipe for a spicy condiment I call Chunky Chile Oil, and developed a process for that product in collaboration with the Food Entrepreneur lab at Cornell."

Michael sold several hundred jars at a local makers market before discovering that his product was vaguely illegal. He put sales on hiatus but continued to grow peppers. He recently gained access to a commercial kitchen and hopes to start legally selling again soon.

This snarl is partly why urban planning experts, like Alfonso Morales at the University of Wisconsin, have advised cities to set up a "one-stop-shop for urban farms, like they have for small business development, so that city farmers can deal with zoning, home business regulations, and nuisance laws all in one place."

In other words, urban farms are generating jobs. They're also recalibrating people's sense of their job options. Urban farming and rural farming are increasingly seen as complements to each other. They are being rightly recognized as part of the same regional food systems. The divide between (city) consumer and (country) producer is also dissolving as knowledge is exchanged. But training is needed.

Back in the late 1990s when Michael Ableman wrote *On Good Land: The Autobiography of an Urban Farm*, he described "the parade of young people" at his door, eager to work on a farm. "Their enthusiasm," he wrote, "tempers on the second or third day when they discover that it is repetitious work, that it can be hot or wet, and that one's mind must be dealt with to stay out there and work all day. Learning how to work physically, tame the chatter of the mind, and push through a sore back or a hot day, takes time and perseverance." This new generation of city farmers is proving they can pull their own weight. As much as city life is about the fast, restless pace, it's also about pushing through difficulty to realize one's potential. Farmer-poet John William Kulm grew up hearing that farmers are optimists, for better and for worse.

And so, to the Bureau of Labor Statistics, we say the future for small, bio-intensive urban farming is strong. As city-born farmer Kristin Kimball describes it, "If you're a medium-sized conventional farmer, things look tough. But if you're diversified with direct-to-consumer sales, then things are pretty hopeful." Kimball credits urban homesteaders in the 1970s "who dropped out to go back to the land—we wouldn't be here without them. Our generation [though] is more focused on maintaining a business out of it and sustaining it in the real world on a long-term basis."

COMMON GOOD CITY FARM

WASHINGTON, DC

In 1984, as part of the District of Columbia Comprehensive Plan Act, the then mayor of Washington, DC, Marion Barry, set out to make better and more systematic use of the district's empty lots and vacant land. The Food Production and Urban Gardens Program stipulated that an inventory should be created and regularly updated, categorizing the location and size of unused lots, and that those lots should be made available to the community for growing food. The new city code further stipulated that DC's urban gardens should be used for education, "programs for citizen gardening and self-help food production efforts," and the creation of jobs.

The Food Production and Urban Gardens Program was a progressive policy move for the time, though implementation did not follow with as much vigor as some might have hoped. It was not until two decades later, as the District and the entire country began to face heightened concern about childhood obesity and diet-related illness, that citizens kicked efforts into high gear and started planting seeds for local, community-driven food production.

One standout urban agriculture project that has sprung up with the new wave of enthusiasm can be found in the LeDroit Park neighborhood of Northwest DC. Common Good City Farm sits on the former site of an elementary school that was closed and slated for demolition when the public school system was restructured in 2008. Members of the neighborhood association advocated forcefully for the land to be spared from development and left available for purposes that serve and benefit the community. They triumphed in their effort to turn the empty space into a park, and Common Good became the anchor in the plan.

The founders of the farm, Susan Ellsworth and Liz Falk, had been running a garden in another location for two years before the LeDroit opportunity arose. When they began in 2007, explains farm manager Spencer Ellsworth (no relation to Susan), "There was a dearth of nonprofits in DC focused on garden education." Though the founders didn't have formal training in agriculture or horticulture, they possessed an abundance of what many people lack: follow-through. "Several of us had worked on farms," says Susan. "And several of us had community development experience through work. But it was mostly started with a vision and an interest in finding a way to use available space productively."

Ellsworth and Falk worked alongside a broader social service organization called Bread for the City. "There wasn't much emphasis in the food pantries and soup kitchens on providing fresh food," Susan recalls. "They wanted to use preserved and canned food because of its shelf life." But over the past several years, that has shifted dramatically, as more people have come to understand the real nutritional and health benefits of fresh produce.

In 2009, Ellsworth and Falk were invited to start Common Good City Farm with the support of the LeDroit Park neighborhood association. The pair recruited friends and volunteers to help them clear the land, build raised beds, and plant the first crops. With support from the USDA, they took soil samples and found that—to their surprise—the land was not contaminated with heavy metals. They scavenged cardboard, newspaper, and hay to start transitioning the land.

The design of the eighteen-thousand-square-foot farm was based on permaculture principles, dictated in part by their programming plans and also by the state of the lot upon their arrival. "When we moved to the new site, we had to contend with a lot of concrete, a baseball diamond made with packed dirt from demolished brick homes. Permaculture design allowed us to create a farm that was resource efficient."

The permaculture design is meant to mimic a local forest ecosystem, with diverse crops planted in close proximity that function to support one another. One of the unique elements of Common Good is their extensive array of fruiting trees, which were donated by a local nursery and include peach, apple, cherry, plum, and a regional native called a pawpaw, which is similar to a papaya.

The majority of the space is run as one unified growing operation, maintained by volunteers and students who share the yield. "We wanted to create a work-trade CSA, where low-income community members could come lend a hand and take away food in exchange for their labor. It was important to us that everyone work together on the whole garden rather than having individual plots where each person is growing tomatoes and basil side by side." As the rest of the land fills out, an area dedicated to individual community plots will also be built for community members who wish to cultivate a small private space and don't have room available where they live, but the collective cultivation remains central to Common Good's vision, particularly with kids, for whom collaboration is a key life skill beyond the garden.

The farm managers direct after-school education programs attended primarily by neighborhood kids. Many of the children live in the public housing complex across the street from the farm and used to go to the school that once occupied the site. Now they walk to schools in adjacent neighborhoods and some are bused longer distances. When they come home at the end of the day, Common Good is a safe space where they can learn about growing and eating healthy food. "We want to instill values and teach skills but without overprogramming," says Spencer. "Kids are forced to sit still and be quiet enough in school, so we want this to be a place where they can play, too."

With the addition of a playground and other recreational space on the same property, the kids have a choice of how they want to expend their excess energy. "It means the ones who do decide to be on the farm are there because they choose it over any old green space." Activities with the young gardeners include hands-on projects that connect familiar food experiences with the unfamiliar process of growing. One of their favorites, reports Spencer, is cultivating and harvesting corn kernels for popcorn. "We put it in a pan, and it becomes what they see in the movie theater," he says.

The kids often go home with some ingredients for their parents to cook and always with lessons about where their food comes from. Occasionally, the parents come across the street to learn alongside their kids and place requests for their favorite fruits and vegetables. The farm managers enjoy these voluntary polls and try to comply, growing extra tomatoes, garlic, collard greens, and fruit. Spencer recounts

an encounter early on in the establishment of Common Good when some young men from the public housing facility came over to tell the farmers about a peach tree that had been growing in their complex for two or three decades, since they were children. They were concerned that new construction near their building might endanger the beloved tree and asked if there was a way to protect or transplant it. Although the tree couldn't be moved, the farmers took the exchange as a good sign of the community's investment in their local food resources, and they planted plenty of peach trees that now fruit in abundance.

In addition to the student volunteers, Common Good has two other active groups of helpers. One comes from a city program called Green Tomorrows, which gives adults who make less than a DC living wage access to fresh food. Green Tomorrows volunteers come from all over the city to dig in at the farm, and at the end of their shift, they walk away with a bag of vegetables. The other group of volunteers represents the growing population of young adult farmers who have affluent backgrounds and college degrees, but choose food production over the more conventional activities that tend to occupy well-off urbanites. "DC is changing pretty drastically," Spencer reflects. "It used to be 74 percent African American, and that is still true in the area surrounding the farm, but soon we'll have a plurality as more Latino, Asian, and white people move in."

As Common Good City Farm becomes more established and diversifies funding sources, they are also building a strong network of partnerships that underscore the many important roles of an urban farm. They get funding from the USDA in the form of community food projects grants that aim to support food security; the Sierra Club provides funds for outdoor education; Kaiser Permanente supports the farm because of its connection to community health and well-being; and DC's Department of the Environment has taken an interest in the potential environmental benefits of the farm, setting up rain gardens there that act as storm water collection demonstration beds.

The one thing Common Good limits is the amount of food they actually sell, so they can focus on their mission, which centers around supporting people who do not have a consistently reliable source of food. "But we are finding ways to balance profit with food security work," says Spencer. They sell herbs to a chocolatier in Philadelphia and select crops to DC restaurants, and they have discussed the possibility of starting a mobile grocery modeled off of projects like People's Grocery in Oakland, California.

Being in DC, Common Good stands near the epicenter of some of the newest movements toward facilitating food access for youth and low-income communities. Michelle Obama's interest in obesity, nutrition, food, and gardening has accelerated the District's progress toward programs and policies that connect schools and food banks with urban agriculture. In 2010, the old 1984 Food Production and Urban Gardens Act was amended with the instatement of the Healthy Schools Act, which is now a working model for the rest of the country. "It's a strong symbol of commitment," says Spencer. "Urban agriculture is an important piece of the overall work of reshaping urban food systems, and we're a city with a lot of opportunity to do that."

GARLIC
STOPS
VAMPIRES
IN
THEIR
TRACKS
ARE
GREAT INFECTION
AND
REDUCE THE
OF PROSTATE
CANCER.

The Herb Spiral
Every plant has different needs. Some like it hot, some cool, some dry, some wet. A spiral hill meets many needs at one time.
In this Herb Spiral the plants near the top like it hot & dry, those at the bottom like it cool & wet.
Dill
Sorrel
Chamomile
Summer Savory
Lemon Balm
Parsley
Chives
Pennyroyal
Mint
I Love Flowers

MMON GO
TY FARM

It's not uncommon to drive around central Detroit and see beautiful old mansions charred by arson and left defenseless to the passing of vandals and time. The city, which in the first half of the twentieth century was a beacon of innovation and industry and a hotbed of new wealth, has now come to represent the economic decline on a great American city.

In 2010, the population of Detroit hovered around 900,000, approximately half of the city's historic peak of 1.8 million in 1950. The unemployment rate was more than 20 percent, and a survey from the previous year showed the median household income at lower than $27,000. While the government developed a project to demolish some of the vacant and condemned properties, it would only impact a small fraction of the more than 25,000 acres of abandoned land throughout the city—more than one quarter of its total area. Plans for what to do with the resulting building material and empty space remained fuzzy.

If there is any silver lining to such a dramatic decline, it might be—as many media reports have said—that Detroit presents an opportunity to experiment with new models for urban living. One of most popular and promising of these new models is farming, which could not only provide more healthy food for city residents, but also generate economic growth and create new jobs. From small-scale backyard farming to multi-acre community operations, urban farming is thriving in Detroit.

FARNSWORTH

DETROIT, MICHIGAN

Most urban food deserts in the United States occupy a small portion of a larger city, spanning low-income areas where nutritious food is scarce and residents lack the means to travel far from the neighborhood to get what they need. In Detroit, however, food deserts are not small islands surrounded by economically diverse and food-rich areas. Virtually the entire city has become what amounts to a food Sahara, with not a single major grocery chain within the city limits. Driving up and down Gratiot Avenue, a central thoroughfare that cuts diagonally across the city, one spots nary a meal option beyond fast food.

But less than a mile off Gratiot, a one-block community of motivated citizens has taken the situation into their own hands, producing food and raising farm animals for themselves and their neighbors. On Farnsworth Street, most of the run-down houses have been bought up either by young Detroit natives who've chosen to stick around and try to implement change or by transplants from bigger cities like San Francisco, New York, and Chicago, who were drawn in by the low cost of living and the creative possibilities of urban decay.

Behind the houses, which are in various states of restoration, the fences dividing the properties have been taken down. The backyards run contiguously, and much of the grass has been replaced with vegetable beds, chicken coops, and pigpens. On one corner of the block, where a house or two might once have stood, Andrew Kemp and Kinga Osz-Kemp maintain a sprawling garden that runs alongside and behind their own home. The beds are dense with vegetables and flowers, wrapping around a dilapidated shed, a few towers of bee boxes, some chickens, and several piles of steaming compost. At the rear of the lot is a small fruit orchard.

The couple has been fixing up their own house for more than a decade, working to pay back the money their friend Paul Weertz paid to purchase it several years before giving it to them. Amazingly, the asking price was a mere $5,000, but Kemp and Osz-Kemp have spent another $50,000 making improvements. Before they took it over, it was a crack house, and then it was used to store bales of hay that Weertz grew to support his own urban farm project at Catherine Ferguson Academy (see page 114), where he teaches biology.

To look at the house now, there's no doubt that the current homeowners intend to bring back the vibrancy of the neighborhood starting with their own abode. The exterior wall facing

the garden is now painted in brilliant shades of yellow, teal, and pink, colors that are reflected in the echinacea flowers, daisies, and abundant green vegetables surrounding it. Turning the garden into the bounty that it is today took work, starting with the quality of the soil Kemp inherited. "Detroit is largely the cheapest fill dirt the city can find," he says. "Before the protocol changed about fifteen years ago, they used to just push houses into the basements and cover them up. In a way the city is full of little landfills. When I dig to plant a tree, I find some pretty ugly stuff."

Soil testing revealed some contaminants and proved very poor growing conditions, but Kemp was committed to remediating the earth around his house no matter how much work it would take. "I am foremost a composter, and then a farmer," he declares, invoking the well-known farmer Joel Salatin of Polyface Farms in Virginia. Though Salatin raises cows, chickens, hogs, rabbits, and lambs, "he calls himself a grass farmer. If we want to garden, we better know how to make healthy soil."

Kemp's garden design decisions were driven primarily by "intuition, inspiration, and improvisation." It is planted in a large circle that functions as a sundial and a henge. "Since I have a wide western exposure, I can follow the sun's course throughout the year," he explains. "This is as important to me as the gardening itself for keeping in touch with the cycles around me." Originally the shape was driven by logistics. "It started out circular because it was most convenient for the tractor to pull the leaves and wood chips around. The circle was the most efficient way to use the tractor." Around the circle, Kemp has dug a trench that collects and stores rainwater during the spring and fall. Eventually he hopes to store water year-round and to raise fish in the trenches.

But this takes time, and the couple keeps plenty busy with day jobs and parenting on top of farming. Kemp teaches English at a local high school, and Osz-Kemp is an artist, whose projects include screenprinting clothing for their small family-owned company, Ink In Bloom. They have two daughters of their own, but to visit Farnsworth is to witness a living example of the philosophy that "it takes a village to raise a child." Kids from up and down the block roam together, wandering through backyards, helping to weed or shovel compost, giving the neighbor's pig a back scratch with a rake, and finding an open door wherever they go.

Around Detroit, Farnsworth has become synonymous with subculture. The residents joke that when they leave the block, it's clear from their hand-altered clothes and the soil under their fingernails that they're "straight out of Farnsworth." Between the home-grown food production, organized skill sharing, and plentiful helping hands among the neighbors, Farnsworth residents hardly have to venture far to get their basic needs met. But their small community is connected to a larger network of citizens around Detroit who are all committed to enlivening the city through art, gardening, cottage industry, and activism.

One of their major civic concerns is the Detroit incinerator, which lies just a few blocks from Farnsworth and sends a noticeably foul odor wafting down the street when the winds are right. The burning of trash is considered by the city to be a sustainable form of energy production, as it generates steam that is used to power buildings downtown. "They call it 'waste to energy,'" says a frustrated Kemp. "I call it 'waste to asthma.'" Garbage is imported to Detroit's incinerator from as far away as Toronto, adding to the air not only the emissions of the burning trash itself, but also the vehicles that haul it hundreds of miles to its final dumping ground. "We need to push the nascent recycling system to starve them of trash to burn," Kemp suggests. "But if they can import it anyway, I don't know. . . . I guess we just continue to remediate in their face."

That Kemp views gardening as a form of revolution is a defining feature of Detroit's urban agricultural scene. It is both reminiscent of the constructive social change movements of the past and reflective of a different kind of future for the city, where the hole left by the auto industry gets filled by another form of productive work—this time the cultivation of local food.

Kitchen Composting

Many people who have never tried composting assume it's messy, smelly, and tedious. For those fortunate enough to live in a city that provides curbside compost pickup service, the barrier to entry is considerably lower, but keeping food scraps in the kitchen can still seem unappealing. There are several ways to combat the odor (and sometimes flies) that accompany compost. One is to purchase a countertop or floor-level composter with a sealed lid and filtered ventilation that keeps everything well contained. Another is to store scraps in a plastic, metal, or glass container in the refrigerator or freezer, where the cool temperatures keep the organic matter from decomposing quickly. If you don't have a garden of your own and your city has no municipal compost arrangement, check to see if a community organization or motivated individual in your area might be offering to take your scraps. Many cities now have bicycle pickup services. Often, community gardens have their own sizable compost piles and welcome contributions from their neighbors. Walking over to drop off your waste may even motivate you to join the garden and get a plot of your own.

Rules around what can and cannot be composted tend to vary, but when making compost for your own garden, it's worthwhile to be particular about what goes into it. Fruits and vegetables, grains, coffee grounds, eggshells, and paper products are all ideal for compost, while meat, dairy, and greasy foods are not conducive to a healthy and low-odor decomposition process. Introducing worms can make for a more efficient composting process, and vermiculture is a surprisingly low-maintenance and low-mess option, even in a small urban home. Transforming the compost into fertilizer requires mixing in nonfood materials such as grass clippings, leaves, sawdust, or shredded paper—this is best done in a large, contained area outdoors. The compost must be turned regularly and covered with a tarp in the event of rain to avoid too much moisture. In a couple of months, all those scraps will be soft, rich, odor-free fertilizer for the garden.

"Gardening and even farming have really increased in Detroit over the past ten years," he says. "I can say that there are far more restaurants, cafés, and breweries giving up their waste for compost, and in turn buying local produce. When I first proposed taking the waste from the Avalon bakery fourteen years ago, it seemed very pioneering and lots needed to be worked out. Now when I ask, there is often someone already taking it."

EARTHWORKS URBAN FARM

DETROIT, MICHIGAN

In the 1890s, Detroit mayor Hazen Pingree established the city's first community vegetable gardens. The initiative, which he called the "Potato Patch Plan," was a response to the growing number of urban poor left struggling and hungry in the wake of a depression that had hit the city. The potato patches were to occupy vacant lots covering more than four hundred acres throughout the city, divided into half-acre plots for the poor to cultivate. By 1896, nearly half of Detroiters who received public assistance were growing vegetables in the potato patches, producing today's equivalent of more than $750,000 worth of food.

Earlier, in 1883, the Capuchin Franciscans, a Roman Catholic ministry, built a monastery in Detroit and began offering social services to needy citizens. In 1929 they established the Capuchin Soup Kitchen, which still exists today and, more than a century after Pingree's Potato Patch Plan, operates one of Detroit's largest urban farms.

Earthworks Urban Farm was established in 1997 by Capuchin Soup Kitchen staff as a small neighborhood vegetable garden on Detroit's east side. The project quickly expanded into a cluster of sites around the neighborhood, all established on vacant lots. The monks at the monastery were concerned not only about their community members having enough food, but about that food being fresh and healthy. "When the Capuchins established the monastery in 1883, they intentionally located it on Mt. Elliott Street, which was then the edge of the city," says Patrick Crouch, farm and program manager for Earthworks. "That way they could have gardens and livestock to provide food for the residents." More than a century later, Earthworks still stands by their guiding mission of providing local food for the local community. Early on they partnered with the Wayne County Department of Health to plug farm-grown components into the public food program and open market stands at local health clinics to make it easier for residents to access and purchase fresh food.

"To us it's very important that our work be viewed within a historical context," Crouch explains. "There's often an idea that what we're doing is brand-new, but the movement—Detroit as an agricultural city—is long-standing, with the ribbon farms (long, narrow parcels of land running along the riverfront) that dominated the landscape and continue to create the texture of the city through the names of the streets, many of which are the names of the original ribbon farmers."

The primary Earthworks site is three-quarters of an acre and includes a greenhouse and a large apiary. Each day, the five staff members and a rotating cast of community volunteers tend the vegetables, haul the harvest to the kitchen, and distribute it through the organization's various outreach programs and through a cooperative buying association that sells produce to local restaurants and markets.

While the layout of the gardens looks fairly standard, Couch explains that he approaches his work with permaculture design principles in mind. "There are no herb spirals or anything like that," he says, "but we are actively thinking about ways to close nutrient loops and turn the farm into a commons." He thinks of it as a design challenge, inspired by the "cradle-to-cradle" concept pioneered by architect Bill McDonough and scientist Michael Braungart, asking how urban waste streams can be redirected to produce food—both literal and metaphorical—for the city.

Crouch came to Earthworks after working on rural farms, where he was troubled by the distance between the place where good food was grown and the communities that needed it most. In coming to Detroit, he sought to close that gap, in part by empowering the community members themselves to produce food where they live. As part of this goal, Earthworks runs a nine-month training program for a small group of individuals from the neighborhood who are interested in learning about and finding work in farming and food production. "The gardens serve as a laboratory and outdoor classroom," says Crouch. "The trainees also visit small-scale processors, distributors, logistics facilities, and restaurants to see the entirety of the urban food system."

During the high season, Earthworks operates a small market on-site, which is run by the members of the training program as a way to learn sales and community outreach skills. As the local food system flourishes and more jobs become available, Crouch wants to see Earthworks trainees taking those jobs and leading the development of their own community, not only through farming but also starting food-related businesses such as restaurants, distribution services, greenhouses, and equipment supply and repair shops. "The gardens are very much metaphors for enlivening the region," he says. "If we can envision what a little piece of property could be, then we should be able to envision what an old factory could be—perhaps a community-run packing house or a lumber salvage operation from decommissioned buildings."

In the future, Crouch hopes that Earthworks Farm's education and training will have been effective enough that the program itself will be rendered unnecessary, or at least take a backseat to the work and leadership emerging from within the community. It's a rare organization that roots for its own obsolescence, but at a deeply mission-driven place such as Earthworks, becoming redundant would be a sure sign of success.

Beekeeping

Many people are petrified by the idea of voluntarily keeping bees near their living quarters. But beekeepers know that a happy swarm is actually quite tame when handled calmly by a confident honey-collector. Bees can be kept in the most urban of environments, including on rooftops, as they will travel up to a mile or more to satisfy their pollen and nectar needs. Keeping the colony peaceful also requires a nearby fresh water source and exposure to warm sun.

Starting an apiary requires a stack of hive bodies—typically wooden boxes that contain frames on which the bees build honeycombs—as well as a brooding chamber. Getting bees into the hive can be done by purchasing an already established colony or by catching a swarm, though the latter is best handled by an experienced apiarist. In some places, volunteer bee-lovers offer swarm management. In Los Angeles, the Backwards Beekeepers set up a hotline and a network of community members who will come collect and resettle swarms when they're causing a problem. Or, in New York, the New York City Beekeepers Association can be called upon for help and advice.

In cities, it's important to select a nonaggressive bee in order to avoid trouble with neighbors and passersby. Many beginner beekeepers start with Italian or Carniolan honey bees, which are both considered to be gentle species. Hives should be installed near high fences or bushes that will require the bees to fly above head-level when entering and exiting the hive. Once the bees are established and making honey, the key piece of equipment (in addition to protective clothing and headwear) is a smoker, which repels and pacifies bees before honey removal. Hive care changes with the seasons, so it's important to understand the bees' needs from the start and to know the signs of disease or parasites. A healthy, happy colony will produce plenty of local honey that can provide a side income or simply add sweetness and immune system–boosting nutrients to your diet.

21

GROWING PUBLIC HEALTH

BY RUPAL SANGHVI

In 1948, the World Health Organization (WHO) laid out the official definition of health as a condition of mental, physical, and social well-being, not just the absence of illness. Given that description, which remains unchanged to this day, even a casual observer of urban agriculture can see how its breadth, scope, reach, and array of activities might influence public health. But it wasn't until the 1996 United Nations International Conference on Human Settlements (Habitat II) in Istanbul that urban agriculture was formally recognized for its contribution to the health and welfare of fast-growing populations in cities around the world. Since then, with the recent resurgence of urban agriculture in the United States, research, practice, and policy continue to expand in these intersecting areas of interest.

Today, as the United States faces some of the worst health outcomes in the developed world, many related to access to healthy food and physical activity, finding creative strategies for effective problem solving is on the minds of many, from policymakers to parents. Policymakers are charged with formulating plans to stem the tide of obesity and other problems related to limited access to healthy food. With almost one out of three young people currently categorized as obese (more than twenty-five million), parents are worrying about issues such as the safety of car seats, which assume a mean weight of under forty pounds for toddlers. Parents are also faced with the discovery that the current generation of children will be the first to have a lower life expectancy than the preceding generation as a result of health issues such as obesity, diabetes, cardiovascular disease, hypertension, and high risk of stroke. These novel concerns reveal the myriad ripple effects of obesity touching day-to-day life, as well as far deeper fissures in the United States' approach to and culture around health, food, and urban design.

Against this backdrop of statistics sloping in the direction of poor health in the United States, urbanites across the cultural, socioeconomic, and age spectra have started to ask themselves deeper questions about food: Where does the city supply come from? What is healthy? What is a sustainable food system? How does food and food preparation relate to health and well-being? Can healthy foods taste good? They are also asking what it is to be healthy, and finding that WHO has had it right all along: Is health just the absence of diabetes, obesity, and heart disease—what about feeling good? How can I achieve a sense of well-being?

Addressing these fundamental questions remains central to advancing health objectives in the United States. In battling an obesity epidemic, public-health professionals have learned that changing individual behavior is the only effective strategy for prevention. However, we have also learned that aiming the "just say no" mantra at unhealthy food choices does not work. It is well known that social environment is a key factor for individual behavior—our patterns neatly align with what our friends, families, and neighbors do.

So how can urban agriculture respond to a health crisis? Across a range of cityscapes, people are picking up shovels and planting seeds with incredible passion and determination, recognizing the potential of urban farming to boost youth empowerment, conservation psychology, and community engagement. Today's farmers are hipsters, hippies, elderly immigrants, youth interns, kindergarteners, prisoners, entrepreneurs, and backyard gardeners. Even food banks are digging in as a way to freshen and diversify the ingredients in their meals.

In the current groundswell, not every urban farming activity or program has been carefully evaluated for its health benefits, but existing studies point to important directions for how farming and gardening can improve health. There is growing evidence that producing food in cities can lead people to seek out healthier food and pursue more physical activity overall. It's also clear that farming fosters social cohesion, mental health, and awareness of environmental issues and the importance of biodiversity. Urban agriculture also provides therapeutic opportunities and creates, by default, a greater number of safe spaces in unstable neighborhoods.

People who participate in urban agriculture are also more likely to prepare their own food and take interest in eating in general. It turns out that practical experience with fresh food—growing, harvesting, identifying varieties, understanding seasonality, cooking, and preserving—positively impacts dietary habits. Several studies show that fruit and vegetable intake, as measured in terms of recommended servings per day, is higher among gardeners than among non-gardeners, and among gardeners versus the average U.S. consumer. Gardening and nature-adventure education in after-school programs also improve the calorie expenditure of twelve-year-olds by 60 percent. And if prepared properly, fresh, natural produce tastes better than the frozen and canned varieties.

There are a variety of nonprofit organizations emerging around the country that are beginning to apply urban agriculture as an intervention to improve the health and well-being of young people. A holistic approach is known to be the most effective way to promote health in the long term as kids grow up, improving not only food choice and obesity outcomes, but also driving decisions around condom use, anger management, and other critical life skills. The programs often employ young people through the summer and expose them to all aspects of food production, from planting to harvesting to the entrepreneurial aspects of selling food, keeping inventories, and managing costs.

Berkeley Youth Alternatives (BYA) is one of the country's more established nonprofits dedicated to helping at-risk youth. Founded more than forty years ago, BYA initiated the Garden Patch on a half-acre railroad crossing lot in West Berkeley in 1993. The intention behind the Garden Patch was to provide multiple-use open space in combination with programs for youth employment. Its components include community garden plots, a children's garden, a youth market garden, and public activities promoting the site as a community center. The BYA strategy is to "help young people develop positive habits that will help them succeed in life."

The presence of vegetable gardens in inner-city neighborhoods also positively correlates with decreases in crime, trash dumping, juvenile delinquency, fires, violent deaths, and mental illness. Violence prevention programs have yet to apply urban agriculture as a technical strategy to reduce violence, but there are promising possibilities in the air.

Innovative prison garden programs strive to improve personal health and mental outlook through pride in nurturing the life of a garden and understanding and connecting nutrition and bodily self-respect. The Rikers Island project in collaboration with the Horticultural Society of New York City is one example of this. GreenHouse, based on Rikers, is dedicated to reducing the recidivism rate by offering incarcerated men and women an innovative jail-to-street program using horticultural therapy as a tool to prepare them for reentry.

The fields of horticulture therapy and conservation psychology demonstrate how human-plant interaction promotes stewardship and induces relaxation, reducing stress, fear, anger, blood pressure, and muscle tension. Working with plants and in the outdoors is known to benefit mental health and personal wellness.

Urban agriculture can also make an important contribution to food access, thereby contributing to community food security. Emergency food providers such as food pantries are now turning to community gardens or starting their own farms. With greater awareness about healthy food, emergency food providers are seeing an increased demand for produce. Urban food production helps reduce food bank outlay to meet the growing demand to procure precisely those foods rarely donated by retailers, restaurants, processors, and other suppliers.

Bed Stuy Campaign Against Hunger is a food pantry that services the Bedford-Stuyvesant, Brownsville, and Ocean Hill neighborhoods of Brooklyn. These areas report the worst health outcomes in New York, coupled with low healthy food access and affordability. Rev. Dr. Melony Samuels, founder and executive director, decided to take it upon herself to improve health outcomes among her food pantry clients. She could see the toll of obesity, diabetes, and hypertension on the clients who are also her community. She began cooking classes and sourcing lean meats. Soon she found demand for healthy food and produce skyrocketing. However, with most people donating canned foods, high-quality produce was hard to come by. She partnered with a local CSA that dropped off unpurchased produce, but soon that was not enough either. So Dr. Samuels began growing vegetables in two lots in order to source her food pantry. More recently, she says she has been tracking health outcomes among her clients and has been able to show a drop in obesity, diabetes, and hypertension.

Since WHO established its definition of health in 1948, the American vernacular around health has changed little. It still focuses on costs of health care, disease burden, prevention of illness, access to insurance, doctors, and hospitals. But with the vitality generated by urban agriculture, there's reason to hope that public health practitioners, policymakers, and urbanites alike will broaden their conceptualization and utilize the important momentum building as a way to create lasting change.

ALL IMAGES FROM VICTORY PROGRAMS
ReVISION URBAN FARM (PAGE 18)

CATHERINE FERGUSON ACADEMY

DETROIT, MICHIGAN

When the Detroit Board of Education told Paul Weertz, a science teacher at Catherine Ferguson Academy, that his students were required to dissect animals as part of their curriculum, he was deeply concerned. It wasn't that Weertz was uncomfortable with dissection—on the contrary, he'd been raising animals and slaughtering his own meat in the city for years—his worry had to do with the formaldehyde the animals are packed in before being sent to the science lab.

The student body at Catherine Ferguson Academy is different from most others in Detroit, consisting of girls from seventh to twelfth grade, many of whom are pregnant, as well as babies and toddlers, who are the children of the teens attending the school. Weertz knew that with pregnant and nursing young women in his classroom, formaldehyde fumes could pose a serious health risk. Nevertheless, graduation requirements were non-negotiable. Weertz decided that the only way to keep the girls and babies safe while fulfilling curriculum requirements would be to raise live animals at the school specifically for dissection (they are euthanized humanely). Such were the first seeds of one of Detroit's most interesting urban farms.

Today, instead of a typical playground and sports field, the outdoor space at Catherine Ferguson Academy houses a menagerie of farm animals and an extensive food-growing operation. While Weertz and several of his colleagues have facilitated the growth of the project, the young students have become the farmers. Their daily schedule includes time in the field weeding, milking, and collecting honey, and even the toddlers take part, spending recess with the chickens, ducks, and rabbits.

For many of the students, as with the majority of American teens, eating fruits and vegetables picked straight from the tree or out of the ground is a novel experience, let alone learning the art of beekeeping. For some, the first encounters with large animals and stinging insects can be frightening, but through the process of discovering they can control a horse or a tractor, the girls also come to realize they can assert control in other areas of their life where they may have felt more helpless. The empowerment they acquire while farming carries over into their relationships at home and often changes their outlook on the future.

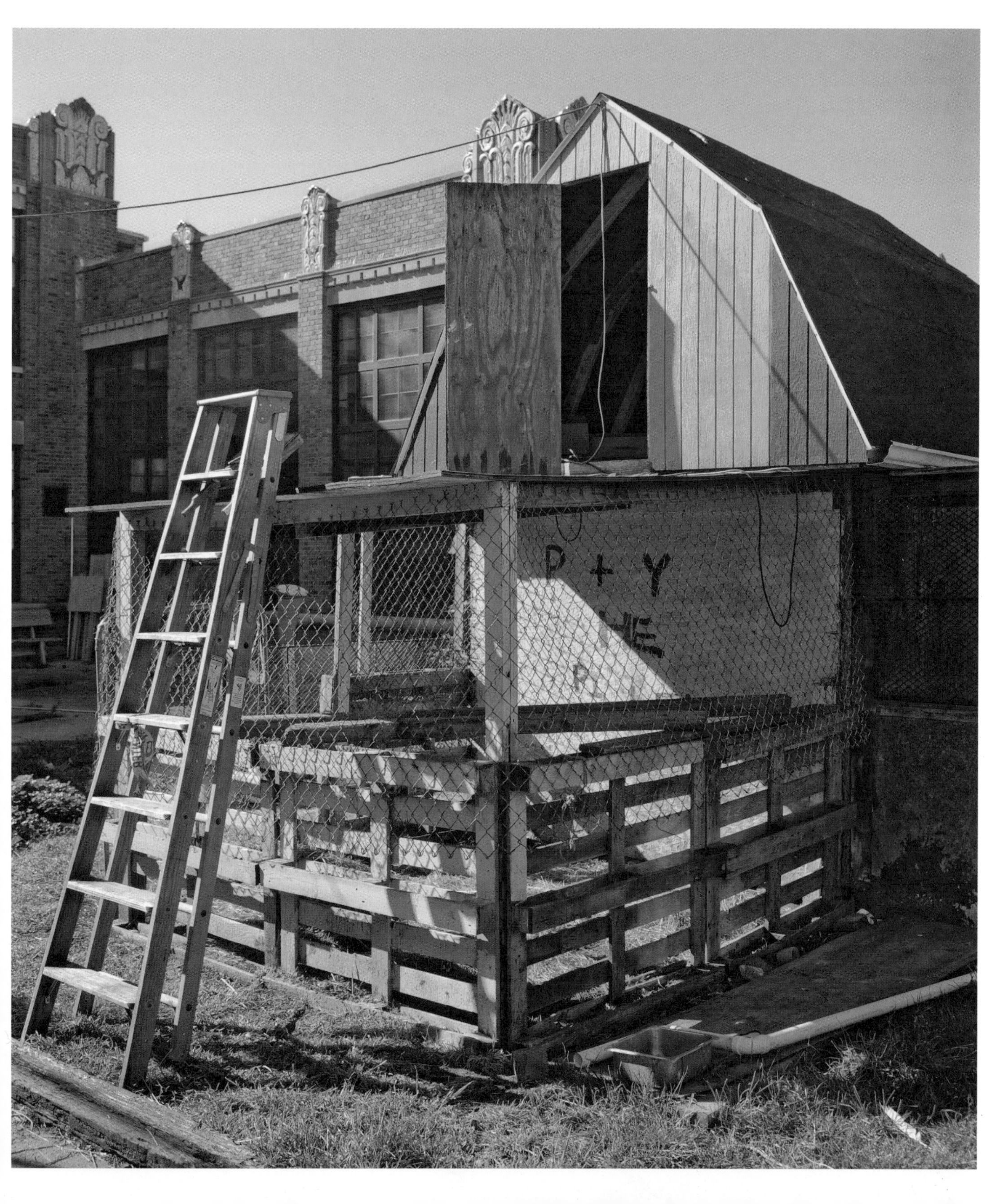
P + Y

LET'S HEAR IT FOR THE
CHICKENS
ORCHARD
ENGLISH
ESPAÑOL
Chicken
Pollo
Egg
Huevo

BLOCKBUSTER
parksideofoldtown.co
ONLY
ONLY

CHICAGO CITY FARM

CHICAGO, ILLINOIS

Many urban farms, no matter how different the cities around them may be, share a common origin story. Most occupy sites that were once vacant and neglected, and in turning those sites back into thriving landscapes, they have contributed to the overall revitalization of surrounding blocks.

According to Ken Dunn, the founder and director of The Resource Center in Chicago, there is a direct connection between vacant land and the condition of urban communities. Simply by making sure that no city lot sits neglected, he suggests, we can ensure better economic stability, safety, community engagement, and quality of life.

Dunn began developing this theory and others while working on a PhD in philosophy at the University of Chicago in the 1960s. "We specifically looked at resources that had been overlooked, such as recyclable trash and vacant lots, and their connection to long-term unemployment." With a combination of evidence and instinct, Dunn decided to create City Farm in 2000 in order to apply his ideas and see what kind of impact local food production could have on the city. The results speak for themselves.

City Farm is situated on the edge of Cabrini-Green, the notorious public housing development on Chicago's North Side. While many of the old housing facilities have been demolished, a relatively new single-resident occupancy building shares a border with City Farm near Clybourn Avenue, and some of the individuals living there partake in the farm's yield.

Designed by German-American architect Helmut Jahn and completed in 2007, the Schiff Residences, as they are called, are strikingly modern, with a five-story glass facade and a set of wind turbines crowning the roof. Between the sprawling, green acre of the farm and the gleaming walls of the SRO, it's hard to drive past this intersection without turning to take another look.

At one time, a gas station occupied the lot, followed by a concrete crushing operation, but when Dunn moved in to start City Farm, the space was empty and overgrown. "The surface was pretty stable," he recalls, adding that soil testing revealed heavy metal contamination. "About twelve feet down you ran into a hydrocarbon problem. We've been doing this for forty years, and we've never found a site that's free of lead, arsenic, zinc, or some other pollutant."

While a complete cleanup is next to impossible, The Resource Center has a time-tested procedure for ensuring that its food grows in healthy soil. The farmers seal the growing area with a six-inch layer of clay, then cover it with compost collected from local restaurants, and wood chips that are available for free around the city.

To look at the farm, raised beds are not visible, but in essence, the entire farm is one raised bed, elevated several feet off of the clay foundation. The wood chips, which are used to form a wide walking path between the beds, add to the sustainability of the garden by absorbing rainfall and minimizing runoff. "The wood chips are porous and the compost is very absorbent," Dunn explains. "The chips hold the water for up to three weeks, so we only use city water to start seedlings. Rainfall in Chicago is adequate enough that we don't need excessive irrigation."

The water-efficient site is ideal for growing crops like heirloom tomatoes, which City Farm cultivates in abundance and sells to Chicago restaurants, along with salad greens and other specialty vegetables. The farm generates enough revenue through restaurant sales to meet a modest overhead, and relationships with local eateries bring value in other forms as well.

Chicago-based celebrity chef Rick Bayless has come to City Farm to work with middle school students, helping them grow ingredients for the Latin-inspired dishes he serves at his restaurants, then working with them to cook meals. "Through working on the farm, kids learn biology, botany, math, and geography," Dunn says, "I think education in our city could be enriched by being more experiential and getting students out of the classroom. It revitalizes education and excites the kids."

City Farm also works with high school–age kids—many from the Chicago Housing Authority—in a more structured arrangement, bringing several students on as apprentices and giving them job training, teaching them skills for earning a living through urban farming and farmers' market operations.

For farm manager Andy Rozendaal, the combination of youth empowerment and agricultural work is an optimal blend of his experiences in church ministry and farming. Rozendaal grew up on a four-hundred-acre corn and soy farm in Iowa, then got a degree in the general agriculture program at Iowa State before attending seminary. After ten years in the church, he was looking for a way to connect his theology background with his desire to address issues of food injustice and urban poverty. "I wasn't finding anything within the church," he recalls. "But when I found The Resource Center, I realized that though it was outside the church, their core values were compatible. This is my dream job."

Rozendaal runs daily operations at City Farm, overseeing the more than eight hundred annual volunteers and visitors, and the small crew of three to four full-time employees. "We don't have electricity, so we do everything by hand, which means we need a lot of help," he explains. The reason the farm is not electrified is not so much a matter of principle as practicality. City Farm does not buy or lease the land for the crops, its use is at the discretion of developers and city officials, with the promise that if the land sells and a building can be erected, the farm will find new accommodations. This fits nicely with Dunn's view of urban agriculture as a treatment for ailing land and communities. The way he sees it, if a developer can successfully sell the land and build new housing and businesses, that's a sign of urban recovery.

Because of this unique arrangement with the city, the farm is designed to be fully mobile. The small office, work shed, and even the wood chips and compost in the ground can be picked up and moved for less than half the cost of establishing a new farm from scratch. The mobile model also supports sustainability, not just in terms of the reuse of farming equipment and materials, but also by ensuring that the city can prioritize density over sprawl.

"I'd like to see all vacant spaces being used as farms," Dunn says. "But you don't want to permanently have farms because then if there's a need for more housing, the alternative is to go outside the city and build suburbs. We don't want to take up areas of the city where infrastructure already exists and force development into virgin areas." In the meantime, of course, they improve the soil and beautify the landscape, so it's a win-win all around.

FARM
ORGANIC
BUY OUR
FOOD!
HERE

STOP
NO PARKING

C

TY FARM

GROWING POWER

MILWAUKEE, WISCONSIN

If you map the Silver Spring neighborhood of Milwaukee online and search for places to eat within a two-mile radius, the results display four fast-food drive-throughs, a fried chicken take-out joint, a Chinese restaurant, and two pizza chains. There's a discount grocery store, a few convenience stores, and then there is Growing Power. In the midst of a typical urban food desert, Growing Power is an oasis of fresh, healthy food.

Nearly two decades ago, Growing Power director Will Allen acquired the two-acre parcel of land on Silver Spring Drive. At the time it was occupied by a run-down plant nursery built in the 1920s, but Allen, who had been raised on a farm, wanted to turn it into a food source for the local community. Today, Growing Power is not only a fully operational working farm, it is also an education center and green jobs resource, and Allen has become a nationally recognized leader in the movement toward urban food security, winning numerous awards including the MacArthur "genius" grant.

Allen grew up on a small farm in Rockville, Maryland, the son of a former sharecropper from South Carolina. While he learned the agricultural ropes as a kid, he was not always a farmer. Towering at six foot seven by the eighth grade, Allen was a natural on the basketball court, and he spent his high school and college years as a serious athlete. Only after moving to Milwaukee with his wife and working in the corporate world for more than a decade did he turn his attention back to food production.

In 1995, at the invitation of the local YWCA, Allen established a nonprofit educational program on the land that would later become Growing Power, working mostly with kids who lived in the nearby Silver Spring government-run housing projects. He and his team restored the greenhouses on the property and began experimenting with techniques that could maximize their limited acreage.

"I was inspired by the integrated farm techniques I'd seen in Europe," says Allen, referring to time he spent in Belgium during his basketball career, where he learned about French intensive gardening strategies that began decades ago outside Paris and spread to the United States in the sixties, and which enable farmers to maximize yield on a minimum amount of land. Both inside the greenhouses and outside, Allen implemented intensive cultivation techniques, including multitier hydroponic systems that produce greens such as watercress, sprouts, and salad greens in the upper levels, and house tanks on the bottom

WILL ALLEN

for raising tilapia and perch. "It's a five-story greenhouse," he explains, though it doesn't look like the skyscraper farms most people imagine when the term "vertical farming" comes up. Growing Power has managed to get all five stories into a single level on a human scale.

It was important for Allen to find a way not only to practice sustainable agriculture, but also to achieve financial sustainability for the farm. Today, in addition to several sizable foundation grants, Growing Power produces and sells enough food to cover about half of its annual expenses, including the salaries of a team of full-time employees. As the staff has multiplied, so have their responsibilities, as the site has come to house bees, goats, rabbits, ducks, and a flourishing flock of hens. Of the creatures they now tend, the one that takes up the most space and brings in the most money is the one least apparent to the eye: worms.

Growing Power's vast vermiculture composting operation yields many tons of high-quality soil, which they sell to other farmers and gardeners in the community. Keeping worms also helps the farm to close the loop on their own operation, composting everything they do not eat or sell back into soil they'll use to grow more produce. The higher-end crops they grow, including lettuces, wheatgrass, and heirloom vegetables are sold to members of a Growing Power CSA, who pay $27 each week to take home a box of organic, local food.

"It's habit-forming," Allen says of the community's interest in farm-grown food. And he isn't just referring to CSA members paying a premium for specialty products. Allen wants to instill in the youth who volunteer and work with him a lifelong commitment to healthy and sustainable urban food systems, starting with what they feed themselves.

In addition to forging relationships with youth and families in the community, Growing Power looks to strengthen the food security ecosystem by involving Milwaukee businesses and public institutions in their work. A local brewery delivers beer mash—the fragrant by-product of their brewing process—by the truckload to add to the farm's Himalayan compost heaps. The mash is combined with shredded coconut fiber and wood chips from the city's tree maintenance services, and eventually shared with the worms.

Each morning at Growing Power, staff and volunteers arrive early, quickly distribute themselves around the two-acre bit of land, and set to work. One of the first tasks is to erect the street-facing market display. Several long wooden tables surround the front door and stretch down the block, laden with giant bushels of produce. Each round basket sits sideways, so that its contents spill decoratively across the table. Electric orange habanero peppers mingle with small purple potatoes, pearlescent onions, green bell peppers, and flowery heads of cabbage.

This cornucopia is one of the first things visitors see upon arrival and acts as a useful attention-grabber for passersby who might not know about the farm. The other thing that's hard to miss, even whizzing down Silver Spring Drive in a car, is the large array of solar panels that double as a sunshade for the market tables and the greenhouse just behind. The panels produce energy for the farm's indoor spaces—the information counter and entrance room, staff offices, and the various pumps and lights that run in the greenhouses. They also stand as a symbol of Growing Power's whole-systems approach to urban sustainability and education. As the name suggests, they aren't just growing food on this farm, they are growing power, which includes the electricity they generate and the empowerment their programs give to the young people who get involved.

Allen likes to call Growing Power an urban farm university, and he consistently seeks new ways to expand their curriculum. In the last few years, they have opened several satellite and partner locations, including a farm in Chicago, run by Allen's daughter, Erika. They have also established training exchanges that allow farmers from around the country to spend time at Growing Power, learning about some of their unique approaches. They teach young farmers to build double-tier aquaponic structures for growing salad greens in concert with cultivating fish, and they've involved visitors in the building of their anaerobic digester, a mechanism for producing energy through the decomposition of farm by-products. Like a university, Growing Power now proudly boasts an expanding roster of alumni who have passed through the greenhouses and garden beds and gone on to start urban agriculture projects of their own in cities across the United States.

8

Will's
Pickling
Cuc's

Pickling Peppers

From sweet Italian peppers to biting-hot Mexican varietals, any kind of pepper can make a fine pickle. Much like canning, making pickles requires boiling water, sterile jars, and fresh, high-quality vegetables. Unlike straight canning, the pickling process infuses the peppers with intense flavor—salty, sweet, and vinegary. Peppers can be canned whole or in sections—when leaving them whole, puncture each one at least once with a knife so they don't burst. In order to avoid a tough skin, the peppers should be peeled by blistering the skin in a skillet or on a grill fire and sliding it off. Combine sugar, salt, vinegar, water, garlic, and any other spices or flavorings you care to add in a pot, and heat them together until a pickling brine has formed. Then, just as in canning, fill the jars with peppers and pickling liquid, and boil them in a pot or pressure canner until they're safely heated. Vacuum-seal the lids and stash the pickles away to spice up a bland winter day.

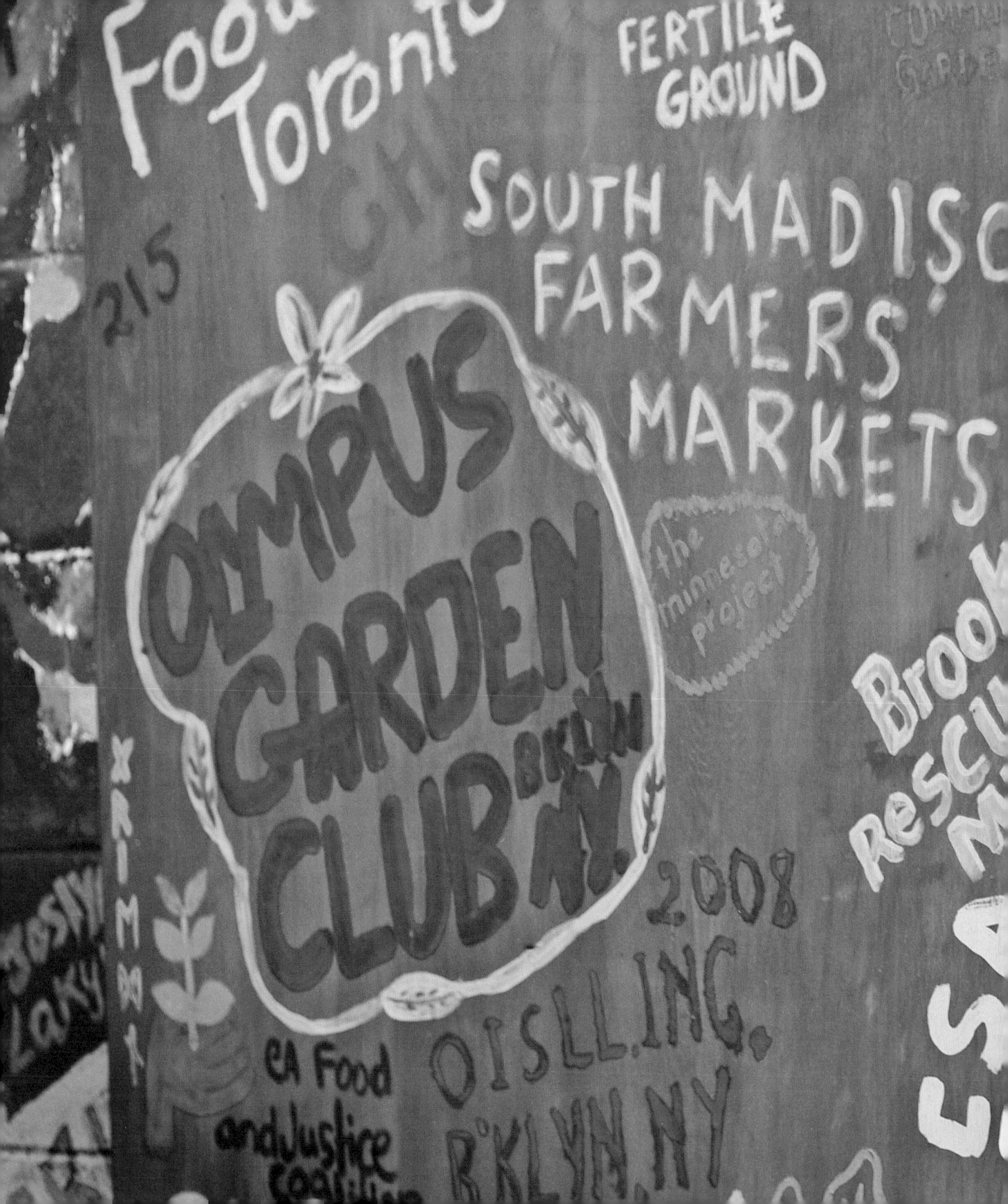

FERTILE
GROUND

SOUTH MADISO
FARMERS
MARKETS

the minnesota project

GARDEN
CLUB B'KLYN N.Y.

2008

OISLL.INC.
B'KLYN.NY

Food
and Justice

PEACE
ROAD

THE ART OF GROWING FOOD

BY NICOLA TWILLEY

Without farmers and farming, cities and civilization as we know them would not exist. Yet for the past ten thousand years, ever since the domestication of grain fueled the growth of the earliest urban settlements of Mesopotamia, agriculture has been outsourced: first in a belt around city walls; later, thanks to improvements in storage and transportation, to a fragmented network of hundreds of different fields, orchards, and pastures around the countryside, and then the world. The distance between urban producers and rural consumers reached its apogee in the early part of the twentieth century, as health concerns forced livestock out of the city and increasing population density put pressure on available land.

The result: By the mid-twentieth century, farming was effectively squeezed—zoned, priced, and sanitized—out of the cities of the developing world.

Yet now, in the first decade of the twenty-first century, with half the world's population living in urban centers, a reverse trend is under way: Agriculture is breaking back into the metropolis, sprouting green on the rooftops, median strips, balconies, and vacant lots of our ever-expanding urban habitat.

In some cases, these examples of urban food cultivation are motivated purely by hunger: the need to grow fresh vegetables in a neighborhood that otherwise might not have access to any, or to supplement the inadequacies of food stamps. In other cases, urban farmers are motivated by a vision that is as much artistic as pragmatic: the idea that growing food in the city is somehow transformative; that inserting agriculture into the concrete and asphalt fabric of the modern metropolis has the potential to change our perception of and relationship to the world in which we live.

Even the most compost-focused community gardener will come to realize that cultivating the city is a practice that inevitably redesigns their relationships with their neighbors and their surroundings. However, many of those who dream of urban cornfields, or actually coax food from the concrete jungle, do so with an explicitly aesthetic, conceptual agenda.

Of course, as with any artistic medium, the use of urban agriculture for creative purposes assumes different forms toward different ends. Take, for example, the work for which American land artist Agnes Denes is best known: *Wheatfield—A Confrontation*. In 1982, Denes planted a two-acre field of wheat in a vacant lot in downtown Manhattan as, she said, a comment on "human values and misplaced priorities." Her thousand-pound harvest was worth just over $250—produced on land valued at $4.5 billion. The grain then traveled to twenty-eight cities around the world under the banner "The International Art Show for the End of World Hunger."

The sight of wheat growing right in the middle of corporate America's beating heart functioned both as a memorial—a monument to the lost landscape of agrarian fertility and sustenance smothered by the glass and steel of New York City—and as a critique of the global capitalist system in which hundreds starve to death in the time it takes for a Wall Street broker to execute a trade. Despite its subsequent harvest and distribution, Denes's *Wheatfield* tilted more toward the symbolic rather than

GREENSGROW FARM (PAGE 58)

the nutritious: a reminder that no matter how fantastic our skyscrapers, we are not civilized unless we have enough to eat.

More than a quarter of a century later, in 2009, American architects Joseph Grima and Jeffrey Johnson planted just 0.2 acres of land in the Chinese city of Shenzhen with grains—wheat, barley, maize—as well as cabbages, bananas, and pasture for livestock. Grima insisted that the installation, entitled *Landgrab City*, "doesn't belong in the category of 'urban farming' proposals." Instead, Grima and Johnson used the field, in the middle of a large and relatively featureless concrete plaza, to create an accurate scale model of how much farmland it will actually take to feed the city by 2027. For every square mile of city, the Shenzhen of 2027 will require twenty-four square miles of agricultural land to fulfill its dietary needs. Despite its bucolic appearance, then, *Landgrab City* might actually dampen the spirits of those urban farmers who cherish idealistic visions of self-sufficiency.

In this example, urban agriculture again functions as a symbol (although Grima suspects that Mr. Yang, the gardener they hired to tend to the plot for the duration of its display as part of the 2009 Hong Kong and Shenzhen Bi-city Biennale of Urbanism/Architecture, consumed the fruits of his labor). However, while Denes's work relied on opposition—the shocking juxtaposition of city and countryside—Grima and Johnson's installation aims to create connection between the two. By growing a field full of food in Shenzhen's busiest shopping district, Grima and Johnson have made the city's ties to the farmland that supports it more tangible than they have been in centuries. In *Landgrab City*, urban agriculture is not about self-sufficiency; instead, it's about uniting urban consumers and rural producers in a relationship of mutual responsibility for their shared foodprint.

Such a shift has the potential to reshape global geopolitics, not to mention the food system itself. It also redefines the city, weaving the metropolis and its agricultural hinterland together into a conceptual unit.

On a smaller, more individual scale, urban farming retains this ability to knit together fragmented space into a reimagined whole. This has been powerfully demonstrated by two recent projects originating in the San Francisco Bay Area: architect Nicholas de Monchaux's 2009 *Local Code: Real Estates* proposal and artist Amy Franceschini's *Victory Gardens* 2007+.

For *Local Code*, de Monchaux used geospatial analysis to identify thousands of publicly owned abandoned sites—first in San Francisco, where the vacant land added up to more than half the surface area of Golden Gate Park; and then in several major U.S. cities, including New York, whose scraps would cover Central Park and Prospect Park combined. He

BROOKLYN GRANGE (PAGE 30)

then reimagined these leftover spaces as a physical urban infrastructure—an "archipelago of opportunity" hidden within the urban fabric—by designing an algorithm that used the constraints of each site to program an appropriate green space function, tailored to local conditions and resources. Not all of the acreage would be devoted to urban agriculture—some might host a swing set or a pollinator park, rather than a vegetable patch—but the cumulative effect would be to give citizens the tools to redefine their city as a distributed network of productive, public space.

Franceschini's *Victory Gardens 2007+* also began in San Francisco and has spread to cities around the world. It also offers urban-dwellers the conceptual and practical tools to transform underutilized space—in this case, mostly backyards, but also rooftops, window boxes, and balconies—into a citywide food production network. In partnership with the mayor's office as well as the San Francisco Museum of Modern Art (SFMOMA), and inspired by the Victory Gardens of World War II, in which Americans grew eight million tons of food per year on twenty million former lawns and parks across the nation, Franceschini's program provides would-be urban gardeners training, tools, and even a shared seed bank.

Franceschini describes the project as "both artwork and democracy in action." She consciously uses aesthetics to reframe farming as an urban, cultural experience, in order to "lure in people who maybe wouldn't have looked at it otherwise, or who moved to the city to get away from this kind of work." And although her project is very much tied into an ethos of local food production as the key to healthy, sustainable cities, she is as interested in cultivating awareness as she is crops. *Victory Gardens 2007+* is, first and foremost, a way to catalyze social and political change through agriculture: It is, Franceschini explains, "another reminder that grassroots efforts engage and mobilize communities and encourage local governments to create change. *This city IS ours!*"

In its activist focus on community empowerment, Franceschini's *Victory Gardens 2007+* fits within a longer tradition of guerrilla gardening. The term was coined amid the crumbling anarchy of 1970s New York, when Liz Christy and her Green Guerillas group illicitly appropriated vacant, trash-infested, privately owned lots in Lower Manhattan and transformed them into vibrant community gardens.

This illegal gardening movement, sometimes edible and sometimes solely ornamental, has taken root all over the world and still flourishes today. It can assume diverse forms, from clay "bombs" studded with wildflower seeds and hurled over a barbed wire fence to carefully tended tomato plants grown on traffic islands, but, in each case, urban cultivation is seen as a radical form of direct action, addressing issues of land use and social justice with the promise of an alternative model of ownership. As Richard Reynolds, who maintains a popular website on the subject, writes: "Guerrilla gardening is a battle for resources, a battle against scarcity of land, environmental abuse, and wasted opportunities. It is also a fight for freedom of expression and for community cohesion."

Growing food in cities is thus deeply entwined with the idea of cultivating a counterculture. Reinserting farming into the urban imagination brings with it the disruptive potential required to renegotiate spatial, political, and interpersonal relationships; to redistribute responsibility, priorities, and resources. It also offers the opportunity to re-situate human activity within a larger ecology and an alternative temporality.

Artist Natalie Jeremijenko's *Farmacy* "dispenses food and food systems that improve environmental health and augment biodiversity" by bringing a cross-species approach to the urban food web. In practice, this means that Jeremijenko expands the scope of agriculture beyond human and spatial relationships in order to redesign our interactions with other city-dwelling species—insects, fish, and even Canada geese—which, like us, rely on and shape the urban ecosystem.

One of her signature projects is the AgBag, a Tyvek sack filled with seeds and soil that drapes over either side of a windowsill or railing. AgBag farmers are able to grow a variety of different crops, including potatoes and tomatoes, in their personal balcony units—but more importantly, they are collectively creating oases of urban grazing for the posse of snails provided with each kit, which, Jeremijenko suggests, are perhaps the most viable source of protein for city dwellers. After all, as she says, given that "snails can navigate the vertical surfaces that make up our urban context" in a way that cows or pigs never could, sustainability-minded locavores ought to consider forgoing burgers and hot-dogs in favor of a tasty escargot kebab.

What makes the AgBag interesting is that it is not just a cheap, parasitic, modular way to take advantage of existing uncontested space; it is also a food production system that ties together two urban, vertical animals—snails and humans—into a new and mutually beneficial relationship. Expanding urban agriculture to the level of urban ecosystems design reveals opportunities to leverage existing assets and design more effective interventions. It also forces its participants to adapt to other, non-anthropocentric timescales: seasonality, annual migrations, and a variety of different growing cycles. Not for nothing is the snail the symbol of the Slow Food movement.

Perhaps even more significant than these underlying rhythms, however, is agriculture's inherent changeability. Edible landscapes are never the same from day to day, or even hour to hour—they sprout, bloom, fruit, wither, and die, sometimes several times each year. Inserting these cycles of rapid growth and equally rapid disappearance into the asphalt, concrete, and brick of contemporary cities is a reminder that although the built environment often seems static, it, too, can continue to change.

A powerful contemporary example of this can be found in San Francisco's Hayes Valley Farm. Built in 2010 on the site of an off-ramp left abandoned after the Loma Prieta earthquake, which prompted the city to demolish the Central Freeway, the 2.2-acre urban farm was only ever intended as an interim use of the land while developers raised financing for construction. But although it is scheduled to be replaced by condominium housing as early as 2012, Hayes Valley Farm will leave behind a strengthened community equipped with food and farming know-how, and—perhaps most important—an expanded sense of both the potential that lies beneath abandoned asphalt and the speed with which the urban landscape can be transformed.

In the end, over and above its ability to supplement people's diets with fresh, local produce, urban agriculture is always an artistic intervention. Whether it takes the form of a plan, a window box, or a field, growing food in the city is a powerful, scalable tool with which to redesign our relationships with each other and with the urban ecosystem we inhabit. It is also, inevitably, a political act that engages with the larger environmental and economic forces that shape today's globalized food system. Agriculture's very centrality to the city's continued existence makes it one of the most potent forces with which to speculate, experiment, and, ultimately, transform urban space.

Havana

OUR SCHOOL AT BLAIR GROCERY

NEW ORLEANS, LOUISIANA

Since Hurricane Katrina, the architectural and economic glue that once held New Orleans together has been weakened in many parts of the city, and destroyed in some. In the Lower Ninth Ward, the strongest remaining bond is social. Most of the families who have stayed have been here for generations. Residents know their neighbors' parents and their grandchildren. Nobody is anonymous.

After school, kids have few safe places to go for recreation or studying. Residential streets—most lacking sidewalks—are the default public gathering spot. But on one corner lot on Benton Street, Ninth Ward youth have a budding alternative, which they are helping to build with their own hands.

Our School at Blair Grocery is a three-year-old urban farm and education center, founded by a Minnesota native named Nat Turner, who found his way to New Orleans by way of New York City. Turner was teaching high school in Brooklyn when Katrina struck the Gulf Coast, and soon after, he began taking his students on an annual trip down south to participate in rebuilding the devastated city. The enthusiasm on both sides of the exchange was so great that Turner established an organization called New York 2 New Orleans, which has been sustaining the long-distance educational trips, even in the years since 2008 when Turner moved away from Brooklyn to set down roots in the Ninth Ward.

The building on Benton Street where Turner now lives and works was once a local grocery store. Owned by a family named Blair, the shop was in operation from the 1950s until 1995, by which time the Blairs' eight children had grown up and moved away. The grocery stood empty from then on up through Katrina, becoming more blighted under forty feet of floodwater, until in 2008, the Blair family made a deal with the newly arrived young man, whom they had come to know during his periodic visits. Turner's connections to the community of the Ninth Ward had deepened over the years he'd been bringing his Brooklyn students, and his desire to effect positive change in the neighborhood could not be satisfied by stopping through only occasionally to contribute to youth education and structural rebuilding. He decided the time had come to make a permanent move. In exchange for a long-term lease on Blair Grocery, Turner promised to fix up the building and the lot. His incentive to improve the building is not only project-oriented, but also personal—he lives on the second floor, where he makes do with rustic conditions as he works his way toward a more fully functional home and school.

When OSBG began, the grocery was too dilapidated to house a classroom—the building lacked basic utilities and a working kitchen, and climbing the rotting staircase to the second floor was a risky endeavor. As a temporary solution, Turner began teaching kids inside a school bus parked in front of the property. But only part of his curriculum involved the students sitting down to listen and take notes. Experiential education is pivotal at OSBG, so Turner wasted no time engaging his new charges in the dirty work.

The ⅔-acre lot was stripped of weeds and garbage, and the stairs inside the grocery were repaired enough to get everyone safely to the second floor, where a makeshift kitchen was installed for Turner to prepare his own meals and students to catch a snack or breakfast. Before long, vegetables were being planted in the yard, and the building's broken windows were replaced with new glass. The Grocery once again became a hub where neighbors met and traded news, and Turner grew into an integral part of that network. Today when visitors come through, he introduces the kids playing out front not only by name, but according to whose grandson or niece they are.

Our School at Blair Grocery is now a thriving urban farm, complete with chickens, ducks, bees, worm bins, and an aquaponic tank full of carp. Every crop, creature, and corner of the farm becomes an opportunity for learning. OSBG began with six full-time students in its first year, and enrollment has been rising ever since. Most of the kids who choose Turner's alternative education program have had problems navigating the public school system or have been released from juvenile detention centers. Some attend conventional school, then come to OSBG in the afternoons, many of them still wearing their school uniforms.

Through the upkeep of the farm, the kids learn biology, horticulture, and algebra. Through the construction of greenhouses and fish tanks, and the renovation of the building, they learn physics, engineering, and calculus. By selling their cilantro and chiles to local taco truck vendors and running the OSBG Sunday farmers' market, they learn economics and Spanish. And, of course, by eating their own homegrown food, they learn not only about health and nutrition, but they also come to understand how delicious healthy food can be. They even learn the social value of preparing and sharing a meal, though most already know that lesson by heart: Communities here have long been woven together through culinary traditions.

In New Orleans, the growing season runs opposite most of the rest of the country. Midsummer months are too hot—for both the crops and the farmers—so the season ramps up toward late fall and runs through the winter. Year-round, a rotating crew of young employees and interns cycles through—many of them from far corners of the country, drawn to the city and OSBG by the desire to gain skills for self-sufficiency while making a positive impact on urban youth. The farm manager, Brennan Dougherty, is one such character, though she's been with the farm since its beginnings and intends to stick around. "I plan to be here at the school as long as is necessary for me to be," she says, adding, "My heart will be here always."

Born in Battle Creek, Michigan, Dougherty earned her agricultural stripes interning on farms up and down the West Coast and in Colorado, before moving to New Orleans post-Katrina. She teaches students and visitors at OSBG to harvest lettuce, collect eggs, and tend to the farm's massive compost pile, which grows daily thanks to food scrap donations from local markets including Whole Foods.

SUSTAINABLE
EMPOWER
MENT

THE AVERAGE MEAL
TRAVELS
MILES
TO THE STORE
FOODS THAT ARE GROWN
SHOP AT FARMERS MARKETS
PLANT A GARDEN

OUR SCHOOL
AT
BLAIR
GROCERY

our school
At
Blair Grocery

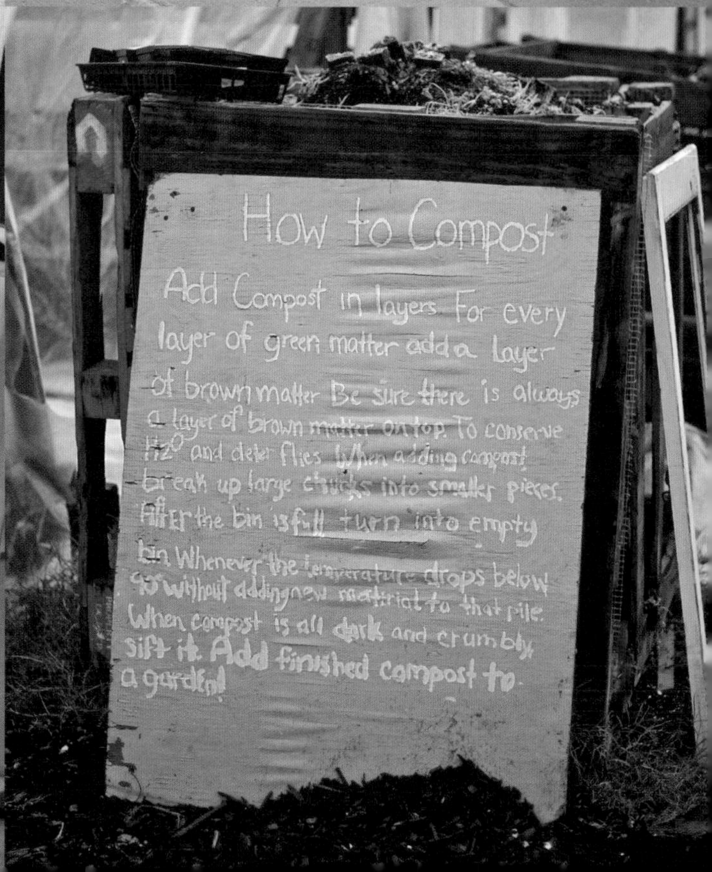

How to Compost
Add Compost in layers. For every
layer of green matter add a layer
of brown matter. Be sure there is always
a layer of brown matter on top. To conserve
H2O and deter flies. When adding compost!
Break up large chunks into smaller pieces.
After the bin is full turn into empty
bin. Whenever the temperature drops below
90 without adding new material to that pile.
When compost is all dark and crumbly
sift it. Add finished compost to
a garden!

markets including Whole Foods. According to Dougherty, the link between the Ninth Ward and the city's more privileged areas is an important part of the OSBG engine. "As we work to provide the means to good food in neighborhoods that struggle the most, we attract guests from neighborhoods more secure than ours, wanting to support our work because of its foundations and also just wanting fresh food."

Our School at Blair Grocery relies in part on foundation grants and private donations to stay afloat, but increasingly they generate a stream of revenue through the sale of their produce. In several handmade hoop houses on one side of the lot, trays of neon green sprouts—mustard, tatsoi, radish, pea, amaranth, beet, chard, arugula, kale, broccoli, bean, and more—prepare for their appearance on the menus of some of the city's best restaurants, including celebrity chef Emeril Lagasse's namesake eatery. Other crops are sold at the Grocery's own weekly farmers' market, as well as through Hollygrove Market and Farm (see page 170), another urban farm and market that has instituted a program to support small-scale food growers around the city. Neighbors are welcome to come onto the OSBG property and harvest produce for themselves, and the garden supplies numerous staples of southern cuisine, including okra, tomatoes, and mustard greens.

As the project has gained traction, OSBG has attracted the attention of national leaders in the urban farm and food justice arenas, including Will Allen, the MacArthur Award–winning founder of Milwaukee's Growing Power (see page 132). Allen now sits on the Blair Grocery board, and many of his farming models have been implemented here—though modified for the climate and ecosystem of the South, where, for example, the low-lying flood plane of the Ninth Ward would not be suitable to invasive tilapia.

While OSBG is already a wonderfully successful example of leveraging urban agriculture for youth empowerment and green jobs, Turner's ultimate goal for the project still sits out on the horizon. "OSBG should act as an academy for food security practitioners within the broader work of sustainable community development across all areas of the green job pyramid," he says. In the long run, he hopes to turn the Blairs' tattered building back into an operating grocery store and to develop the surrounding block into a vibrant stretch of locally owned businesses that provide necessary services and employment—and homegrown entertainment—to anchor residents in the neighborhood.

Until that dream is realized, the young farmers are maximizing their productive land and even moving across the street and around the corner, planting food in other abandoned lots. One neighborhood woman, who used to pick the okra at OSBG as frequently as possible, now has an okra garden all her own, planted by the students.

Eventually Turner and Dougherty hope to extend the learning all the way from seed to skillet by installing a fully operational kitchen and teaching the kids to cook what they grow. In preparation, they encourage the students to eat—and learn to enjoy—some of the plants in the garden. While they still frequently opt for sugary soft drinks over green vegetables, Dougherty says, "each kid usually finds at least one vegetable they like a whole lot." Some of the crops with which they're least familiar, like carrots, grapes, strawberries, and fragrant herbs, often become the favorites. "Maybe that's because of that desire kids have for discovery itself," she muses. "It's beautiful, whether it makes a full meal or not."

MONKEY

Havana
Star Ship Nine

Blair
Fresh
Vegetables
0 lb
8 oz
1
2
3
4
5
6
P&D
PAULA DEEN
Max 6lb 8oz

MONKEY

Havana

HOLLYGROVE MARKET AND FARM

NEW ORLEANS, LOUISIANA

The entrance to Hollygrove Market and Farm in New Orleans doesn't look like your typical urban farm. Instead of the haphazard, utilitarian structures that often occupy agricultural lots, Hollygrove's walkway is lined with a set of clean, modern pavilions, which lead to a marigold-yellow house with red-rimmed windows. The structures were designed by architecture students at Tulane University, and their orderly appearance reflects the organization's mission to provide mentorship and support to local food growers who need assistance getting their produce to market.

The polished property is also a symbol of recovery in a neighborhood hit hard by Hurricane Katrina. The Hollygrove and adjoining Carrollton neighborhoods were both severely flooded and subsequently lost many residents. But this area was also one of the earliest to bounce back, driven in part by motivated cadres of local volunteers, as well as small grants from charitable foundations.

The Hollygrove Market and Farm was part of that rebuilding, constructed in 2008 on the site of a former nursery that was totaled in the storm.

Founded by Paul Baricos and Kevin Fitzwilliam, the original impetus for the project was to serve nearby residents with affordable, sustainable produce grown within the neighborhood. The added benefit, of course, was economic—the farm was a local revenue generator not only for Hollygrove Market itself, but also for individuals who aspired to grow and sell their own food but needed the financial safety net of a supporting organization.

Hollygrove Market and Farm was developed through a collaboration between the Carrollton-Hollygrove Community Development Corporation and the New Orleans Food and Farm Network (NOFFN), a local nonprofit that supports numerous small-scale agricultural projects citywide. One of the first things NOFFN does when helping to establish a new farm is to test the property's soil for lead and other contaminants. Ariel Wallick Dorfman, the urban agriculture educator at NOFFN and an active Hollygrove farmer, says that after Katrina, there were elevated concerns around hazardous ground pollution, but for the most part, lead levels were equivalent before and after the storm—often too high to safely grow food directly in the ground.

At Hollygrove, raised beds were built on about half of the land, where soil tests showed unsafe pollution levels. With about twenty small individual plots, this side of the farm is

designated for individual community farmers who might be learning to grow food for the first time in a relatively small and manageable area. On the other side, where the soil was not contaminated, long beds run at ground level, tended by just two farmers who earn their livelihoods from their harvest.

One of these loamy production plots is cultivated by a local character affectionately known as "Macon Fry, the Lettuce Guy." Fry is a veteran gardener whose vibrant crop of green and red lettuces has become a fixture on many restaurant menus around town. The weathered, almost waifish southerner can be found every day squatting beside his beds, deftly beheading lettuce plants and nestling them into boxes for delivery.

After several decades teaching in the New Orleans public school system, Fry turned his attention exclusively to growing food and discovered that "if you're willing to work hard enough, you will succeed. Farming is an avocation that yields to labor." In this way, he adds, "it is the opposite of public school education." But while he left the classroom behind, Fry continues to be an educator. Ask any of the young farmers throughout New Orleans who has inspired and guided them, and Fry's name will come up.

Because there are full-time farmers working at Hollygrove, the site is always active, but it comes alive on market days, when the main building opens its doors to shoppers. The market, which is accessible twice a week, sells vegetables grown on-site by members of the community, as well as produce grown around New Orleans by other small farm organizations and individuals. The market manager's sourcing radius also extends outside the city, partly to support nearby rural farmers but also to round out and diversify the market's year-round offerings. Hollygrove also runs several smaller, off-site markets around town, including one at Tulane and another in the French Quarter, where tourists have a chance to see local agriculture in action.

Hollygrove's business model is one of its most unique characteristics. The market manager purchases inventory from the farmers up front, then sells it at a markup to the public, absorbing any financial burden that comes back on unsold goods. For the individual growers, this service provides financial security that they wouldn't have by selling directly to consumers and removes some of the logistical demands and time commitment generally required to run a market stall. Though consumers pay a slight premium, they gain the advantage of having a single, well-maintained, indoor destination where they can always expect a wide variety of fresh, locally grown produce.

TOMAT

Canning Your Crop

In order to store a significant food reserve these days, a lot of people have extra freezers, a second refrigerator, and a pantry filled with dry goods that contain preservatives sufficient to extend shelf life by months or years. But before refrigeration and industrial food production, canning was the primary way to ensure there was enough to eat—and a plentiful variety—during winter months and throughout the year.

The trick to canning food that will remain safely edible for an extended period of time is making sure that the preparation process is sterile and hot. The acidity of the food being canned determines how high the heat must be in order to kill dangerous microorganisms and what method of canning to use. For low-acid foods—most vegetables—a pressure canner is recommended to apply heat nearing 250 degrees Fahrenheit in a pressurized environment. High-acid foods, which include most fruits, can be processed in boiling water. It's critical that the food is very clean, the jars sterile, and the lids sealed tightly—all of which guarantee the safety of the food and also help maintain the rich, natural color of whatever is being preserved.

LITTLE HOMESTEAD IN THE CITY

BY ALISSA WALKER

OPPOSITE: FARNSWORTH (PAGE 92)
PAGE 181: GHOST TOWN (PAGE 192)

Straddling the border of the Los Angeles neighborhoods Echo Park and Silver Lake, just a few miles from downtown, the front yard of Kelly Coyne and Erik Knutzen's home stands out in stark contrast to the neatly clipped lawns of the 1920s-era houses on the block. Their front yard swirls with edibles that tumble over the fence and down toward the sidewalk, from fluorescent orange nasturtiums and towering prickly pears to frilly lavender and tiny green tomatoes to hops that will eventually ferment into a deep golden lager. In back, chickens strut past artichoke plants almost as tall as the house, and bees hum from a hive that looks like a cheerful yellow chest of drawers. The garden is strewn with clever DIY experiments: a solar oven, a wood-burning stove, a not-quite-as-sexy hack on their washing machine output valve that dumps gray water onto their citrus trees. Even a stroll through their pantry, stacked high with mason jars cradling handmade spicy pickles and homemade flaxseed salve, is like visiting a museum showcasing the beauty and practicality of living off the land—and all less than a mile from the 101 freeway.

For thirteen years Coyne and Knutzen have been at this address, cultivating a new kind of urban lifestyle. As the authors of the 2008 book *The Urban Homestead: Your Guide to Self-sufficient Living in the Heart of the City* and the popular blog Root Simple, the duo are two of the best-known champions of the burgeoning urban homesteader movement. But while the first incarnation of the phrase, which hails back to the 1970s, was more about a radical, politically fueled self-sufficiency, Coyne and Knutzen see contemporary homesteading as a way to save money, work with their hands—and have fun. "Why should we limit where this movement goes by having a dour 'sustainability' message?" asks Knutzen. "Right, left, rich, poor, theists, and atheists all care where their food comes from and the quality of life for future generations."

Aimee McAdams shared a similar sentiment when she started her homestead in Minneapolis. Although she'd always been interested in gardening and sewing, the move to a new city gave her a chance to act upon her and her husband's fantasies to eat only what was grown within a hundred-mile radius, and most of it from their own backyard. The garden they started became a gateway to more challenging tasks like preserving vegetables, building a root cellar, and growing mushrooms. "Once you start in on one or two things, it just goes from there," she says. "I think of it as a process, not a defined state." She launched her Adventures in Urban Homesteading blog to chronicle her triumphs and challenges, as she experiments with adding new skills.

Running a farm in a dense environment definitely means being the quirky house on the block, says McAdams. "They think of us as having a very different lifestyle." But McAdams has happily accepted her role as ambassador by giving tours and sharing produce, realizing that simply letting people experience the homestead for themselves can transform that skepticism into a contagious sentiment about what they're doing. She remembers as they were building their chicken coop, fielding less-than-enthusiastic queries from neighbors

who didn't like the idea of living next door to chickens. "A week after they had moved in, one woman said, 'When are you getting those chickens?' She didn't even know they were there," laughs McAdams. "Now she loves them, and she's over there cheering them on."

But neighbors don't always come around. Someone snitched on McAdams's husband's mushroom farms—large logs spread with oyster mushroom spawn and wrapped in black bags—and they got cited by the city. They first distributed the mushrooms to several neighbors' yards in an attempt to diffuse the situation, but eventually they had to move the logs to a plot of land in a more hospitable part of town.

Whether willingly or not, urban homesteaders are often finding themselves at the forefront of legislative battles to change zoning laws. When James Bertini, who runs the local agriculture center Denver Urban Homesteading, acquired five hens for his South Denver property two years ago, he also received a thorough education in his city's local politics. Archaic laws—set in place in the 1960s when Denver was trying to clean up its Wild West image—required a $150-per-year fee and a ridiculously overwrought process. "To get a permit was difficult, expensive, and time-consuming," he says. He launched the site FreetheChickens.com and began working with city council members to change the law to allow up to six chickens without a permit—a campaign that has lasted more than two years.

Still, it's frustrating for Bertini to work with politicians who can't see the big picture of how an urban homestead is supposed to function. A wrinkle in the campaign occurred when the proposed new legislation wanted to outlaw the slaughtering of chickens—something Bertini sees as intrinsic to the homesteader lifestyle since most chicken owners want to raise and eat their own meat. But those concepts don't register with the city council members, who only see the chickens as backyard pets. "They don't understand that people want to live more sustainable lives," he says. "They just have heard that people want eggs." To promote better understanding, Bertini organizes chicken sales and chicken coop tours, and offers a chicken recycling program that will humanely slaughter a bird for homesteaders who cannot do so themselves.

Motivated by his chicken activism as well as coaxing from his wife, Irina, who hails from Turkmenistan in the former Soviet Union and was raised on unprocessed foods, Bertini opened Denver Urban Homesteading as an education center and farmers' market. Classes on building cold frames and canning are taught by local experts, and an indoor market features organic and artisanal foods from sauerkraut to raw milk, all from small, local purveyors. The hope is that the center can create awareness about industrial food production and inspire people to make their own changes. "People like the fact that they can get some level of education at the place where they're getting food," he says. "Even if they're not planning to take classes, they like the fact that it's there."

If a resurgence in urban homesteading means a renewed interest in pioneering skills, it doesn't come with a renunciation of technology. Urban homesteaders trade tips about growing tender asparagus on Twitter and share photos of their impressive honeycombs on Facebook. Online dialogue can explode when debating the merits of various composting methods (just Google "lasagna gardening" to see the heated arguments for layering cardboard and kitchen scraps into a simple self-composting bed versus a more rigorous and tidy bin composting method). Especially effective for community-supported agriculture shares and local markets where seasonal offerings change each week, a Facebook or Yahoo community can facilitate a vibrant discussion around sharing, growing, or preparing local foods. And it also helps customers get more information and even see photos about where their food comes from. "I update our Facebook page several times a week about food issues at the market," says Bertini. "It's a really good way to tell our customers what we're doing."

Online communities have also become a way to catalyze local grassroots movements around homesteading issues. Knutzen launched the Los Angeles Bread Bakers Facebook community after he discovered a shared obsession with a few bread-baking friends. The group trades recipes, buys flour in bulk, and plans community oven-building events. Another group that Knutzen is affiliated with, Backwards Beekeepers, helps to "rescue" unwanted hives—they even have a Bee Rescue Hotline—from homes and place them in more welcoming backyards. An online Yahoo discussion group and how-to videos keep the larger community informed, while regular in-person meetups allow the group to demystify the beekeeping process for newbies.

Urban homesteaders reach an even larger, global audience through their blogs, where succulent photos of homemade goat cheese are posted among the wiggly trials of vermiculture. In a larger way, the practice of blogging has taken the place of the advice that would have passed via word of mouth, from generation to generation. Coyne and Knutzen have written and spoken extensively about this culture of sharing. "I think we have a responsibility, especially since we live in a media capital, to speak and help spread the word," says Knutzen.

Sharing also extends to a whole range of goods and services that urban homesteaders are producing. Whether it's holding informal produce swaps with neighbors, founding time banks to register local skills, or organizing tool libraries, the move toward self-sufficiency also means finding what you need in the homes around you. "Bartering has been more and more important to us over the last few years," says McAdams. "Part of that has been discovering the skills of the people around us." McAdams offers her sewing skills and her husband's homegrown mushrooms in exchange for electrician and construction skills found in blocks around them. Now McAdams is creating a hyperlocal directory of services for their fifteen-square-block neighborhood which will include everything from brick-and-mortar businesses to people who want to trade their surplus of apples.

With the introduction of specialized skills and the creation of localized market economies, the homesteader movement is now evolving beyond the traditional crops-and-chickens model. McAdams launched a new blog, From the Homestead, just to showcase her sewing projects, like making pillows from vintage ties, which she sells at local markets. Bertini's Denver Urban Homesteading classes include workshops on how to restore wood furniture. And Coyne and Knutzen's second book is entitled *Making It: Radical Home Ec for a Post-Consumer World*. In their book, Coyne and Knutzen chose to include recipes for making homemade kimchi alongside directions for making soap. For them, broadening the definition of homesteading widens the point of access to encompass eager do-it-yourselfers who might not think of themselves as farmers. "One of the things I like about 'home ec' is that I think it's less off-putting to folks living in condos and apartments," says Knutzen. "There are plenty of activities generally described as 'home ec' that can be practiced without access to land."

So if life in the city is better with a backyard beer factory, a basement incubating mushrooms, a few chickens, and maybe a goat, then the question remains: Why not just move to the country? Urban homesteaders say even with all the challenges that a tiny backyard plot presents, it's this dense environment that makes their operations work. Bertini cites the lifestyle of being able to experience a little slice of pastoral life, yet still be able to hold down a lucrative, high-tech job—the best of both worlds. For Coyne and Knutzen, it's all about the tight-knit community that comes from living in close quarters with their fellow Angelenos. And for McAdams and her husband, the city gives them the best platform to share their skills. "There's so much opportunity for educating people around us," says McAdams. "We want to show people that this isn't totally crazy."

NEW ORLEANS HOMESTEAD

NEW ORLEANS, LOUISIANA

When Jordan Shay and her boyfriend, Albert Walsh, bought a house in the Freret neighborhood of New Orleans in 2009, the building was in grave disrepair, and the block was much the same. Even before the storm, says Shay, "It had been seriously neglected for years. It sat on the foundation crooked, and the two back rooms were full of rot and unsafe to walk on. It was by far the worst house on the block and a total eyesore." This uptown neighborhood, situated just a short distance from Tulane University, did not suffer as much destruction during Katrina as some other parts of town, but the depressed economy and reduced population contributed to keeping an already run-down neighborhood from turning itself around quickly. Shay and Walsh were part of a small influx of new homeowners intent on restoring the housing stock and opening some businesses that would serve and revitalize the area.

While Walsh, a builder who specializes in historic renovations, worked on shoring up the structure itself, Shay set to work in the backyard, turning a field of weeds into a food source. After testing the soil and discovering that it contained high levels of arsenic and lead, she laid down compost and mulch, planted sunflowers to draw the lead out of the ground, and purchased ten tons of topsoil. "The first season I only planted fruiting vegetables—tomatoes, squash, peppers, eggplant, okra, beans, corn, watermelon," she says, "because plants only absorb lead in their stems, roots, and leaves. Plants like herbs, lettuce, and root vegetables I planted in raised beds just from imported soil, free of lead."

Growing a variety of vegetables is not unusual in a town with such a garden-friendly climate, but Shay distinguished herself from other neighborhood green thumbs when she brought bees, rabbits, chickens, ducks, and pigs into the picture. At that point her quiet garden became a full-on farm.

A native of Santa Cruz, California, Shay has a background in agriculture. She studied agroecology at the University of California, Santa Cruz, and then applied her skills at various farming jobs from northern California to New Zealand. In 2005, after Katrina hit, she traveled to New Orleans to participate in the rebuilding effort by offering her landscaping and horticultural skills to local residents, finishing her degree concurrently at Loyola University New Orleans. "I never expected to stay," she reflects. "I think a big part of what has kept me here is that it is a city where I can be as social as I want and surround myself with people my own age, but I can also still maintain a connection to animals and grow my own food."

She also embraces the relative freedom afforded by the city's urgent need to rebuild, which has meant that determination is almost all a person requires in order to start farming a residential plot. The city rarely interferes in activities that are a net positive for a neighborhood, and Shay found that this even extended to squatting on other people's vacant land—at least for a while.

The lot adjacent to Shay's had sat empty for several years when she and Walsh moved in, so she went ahead and planted food crops on the other side of the low fence dividing the properties. She even experimented with a few exotics—such as bananas and papayas—which can eke out a few fruits in the humid southern air. After a time, the owner of the neighboring property, who was having trouble selling off the fallow land, decided perhaps Shay ought to pay for the productive plot on which she was squatting. Not wanting to increase her overhead, she pulled the operation and allowed the vegetables to grow back into weeds. The chickens and ducks, however, still roam freely across both lots.

In a city with such an abundance of homes in need of repair, Shay and Walsh have been on a cycle of moving from one house to the next, fixing them up as they go. This house was no exception—they moved out in 2010—but Shay had such a well-established farm at that point that when it came time to rent the place out, she looked for tenants who would take a liking to her livestock. She got lucky with a pair of renters who were charmed by the idea of walking down the back stairs to find ducks fluffing their wings in the small pond and pigs cooling themselves in the mud.

Though Shay no longer lives on the property, she rides her bike over daily to tend to her flocks, weed the beds, and harvest what's ready for eating. For the most part, she does not sell her yields, preferring instead to feed herself and Walsh, and to share with her community. Much of what she grows goes straight to the plate, but in her ongoing experiential education as a homesteader, she has learned to make value-added products including pickles, preserves, strawberry wine, and honey mead. "I also make soap from the lard rendered from my pigs and beeswax from my hives," she says. "It is a rich and foamy soap, but I have to add essential oils like chamomile or lavender to get rid of the slight fatty smell."

She has also learned to slaughter her animals, though after raising, killing, and butchering a few herds of rabbits, she decided they weren't for her. "I hated slaughtering rabbits," she admits. "It just never seemed to go smoothly for me. I liked the animals themselves, but I never felt right about having to keep them in cages. Every other animal I've raised for meat has gotten to graze and forage."

Compounding her own discomfort with the act, she received hateful and threatening comments after writing about it on the blog she keeps about her farming exploits. While she still keeps pigs, chickens, and ducks, her rabbit cages have been empty for some time.

Like every urban homesteader, Shay is learning as she goes, finding what works best not only for the climate and in the context of her neighborhood, but also what suits her best as a farmer. While not too many locals run residential operations as extensive as hers, she sees quite a few members of the community learning to grow food—many of them motivated by how few grocery stores reopened in the aftermath of Katrina. For reasons of economics and convenience, the city certainly needs its food markets to make a comeback, but if their scarcity prompts citizens to become a bit more self-sufficient, it's a beneficial side effect for New Orleans.

Housing Chickens

The only thing as uniquely delicious as a homegrown tomato is an egg freshly laid right outside your door. In the last few years, cities have been jumping on the chicken bandwagon, allowing residents to keep coops in their backyards in limited numbers. It's important to heed regulations (no crow-at-dawn roosters, for example), but once your hens are situated, maintenance is easy and the rewards are many.

A DIY chicken coop can be constructed with basic building supplies, but if you're more of a design connoisseur, there are numerous commercially available prefabricated coops, some of which may be more sleekly modern than your own house. Because coops can be a small, manageable size, many novice designers take an interest in creating innovative chicken coops as a platform for small-scale experimentation.

No matter how you build your coop, be sure it has a completely enclosed area with a roof and a small doorway just large enough for an adult chicken to get through. Coop dimensions should provide at least three square feet of space per chicken, and it's important to avoid overcrowding, not only to raise happy chickens, but to ensure regular egg-laying. The coop should include a secure roosting bar where birds can perch comfortably side by side and a box for laying eggs. There should also be a sufficient enclosed area outside the coop to let the chickens roam in the open air. A run of four to five square feet per bird is recommended. At night, chickens must be locked safely away from the threat of raccoons, dogs, and other predators.

The range of species kept in residential backyards is fairly vast, though common choices include Ameraucanas, Leghorns, and Minorcas, all of which are considered to be prolific egg layers, at a rate of at least four eggs per week. Ameraucanas and Araucanas are often nicknamed "Easter Eggers" because of their colorful blue eggs.

Chickens will eat a range of vegetarian feed, including kitchen scraps, but standard chicken feed is an easy choice and can be purchased at most pet stores. Before you order that box of fuzzy chicks, be sure to do your research and be prepared for caretaking that's a degree more demanding than raising a dog or cat.

28
Martin Luther King Jr
STOP

GHOST TOWN FARM

OAKLAND, CALIFORNIA

In late 2009, Novella Carpenter traveled to Brooklyn to teach a workshop on butchering rabbits. Participants paid $100 each and went home with the main ingredient for a fine meal. The *New York Times* ran a long feature on the event, accompanied by recipes for rabbit ragù and rabbit loin with bitter greens. Afterward, Carpenter flew back to her hometown of Oakland, California, with the satisfaction of having trained a few more local food loyalists in the hard-core art of butchery.

It didn't take long, though, for the pleasant memory of her trip to be clouded by hate mail, which poured in from vegetarians and animal rights activists, who were outraged at the thought of urbanites slaughtering rabbits as a weekend diversion. What the commenters and letter writers may not have realized is that for Carpenter, this is not a hobby. On days when she's too tired for hard labor, she doesn't simply drive to Whole Foods and purchase meat for dinner. She tends daily to rabbits, goats, chickens, ducks, bees, and at one time, a pair of hefty pigs. At the end of the day, she cooks food she has raised or grown. Carpenter is a farmer—she lives off her land.

A tall, blonde thirty-something whose style could be described as utilitarian-vintage, Carpenter talks about her occupation with blunt—sometimes raw—realism. She doesn't put a gloss of romance on her experience, though it often gets applied from the outside, as urbanites increasingly aspire to get a little dirt under their fingernails. In late 2010, she was featured in *Vogue* magazine by a writer who temporarily immersed himself in her lifestyle and managed to describe even the most malodorous moments with a touch of glamour.

Carpenter calls her operation Ghost Town Farm, after the condition of the West Oakland neighborhood in which it sits. These days, the ghosts of neglect have surrendered, to some extent, to the controversial advances of gentrification, but when she first arrived with her boyfriend, Bill Jacobs, in 2002, the area was largely composed of abandoned buildings and lots, and populated by drug dealers, drug users, and frequently clashing gangs.

Carpenter and Jacobs didn't come to the neighborhood as typical gentrifiers. If someone were to open an upscale coffee shop nearby, the couple would not come in for lattes, though they might be found out back after closing, rummaging through the shop's dumpsters for uneaten pastries to feed their livestock or coffee grounds for their compost pile.

Carpenter's agricultural proclivities trace back to her childhood in Idaho, where she was raised by back-to-the-land hippies; and through her teenage and college years in Washington state, where in many cities backyard chickens are common. When she landed in Oakland, her new environment was far less bucolic than most she'd known, yet it was perhaps the most laissez-faire place she'd found for establishing a farm. With crime consuming the attention of law enforcement, Carpenter's pursuit went relatively unnoticed. At least at first.

On the top level of a duplex on a dead-end street, Carpenter began her project with a stack of bee boxes. She kept them on the deck off her living room in order to avoid frightening any skittish neighbors. Next came mail-order chickens, ducks, and turkeys, which were housed inside the apartment while she built a coop and strengthened the young birds enough to cope on their own on the ground.

While the house Carpenter and Jacobs had moved into didn't come with abundant yard space, the adjacent lot was vacant—a 4,500-square-foot swath of weeds and concrete where a house had once stood. After testing the soil and finding it safe, Carpenter got to work planting almost every square foot with something edible: cucumbers, onions, carrots, garlic, tomatoes, squash, pumpkin, artichokes, herbs, beans, berries, and an array of fruit trees including fig and citrus, which thrive in the Bay Area's Mediterranean climate.

She took her first leap from feathers to fur when she brought home a small colony of rabbits, for whom she also made a home close to her own sleeping quarters, elevated from street-level dangers. Through steadfast self-education, Carpenter learned to breed and butcher her small meat animals, and through instinct, she learned to defend them when the occasional predator did find a way onto her property. As the farm grew, neighbors began to notice the addition of agriculture to their gritty environs, and Carpenter forged relationships with a cast of characters fit for a novel, from the homeless man who lived out front in a broken-down car to the monks who occupied a quiet monastery across the street to the Berkeley hippies who biked over from their idyll up north to check out Carpenter's experiment in personal sustainability. She did, in fact, later chronicle her experiences in an autobiographical book titled *Farm City: The Education of an Urban Farmer*, published in 2009.

After the successful slaughter of her first homegrown livestock—and the delicious rewards of their meat—Carpenter gained the confidence to scale up her animal kingdom. Behind her house she built a pen and raised two pigs, simultaneously apprenticing herself to an Italian salumi chef in order to prepare the hogs for butchering and cure their meat. Raising pigs on an urban lot proved challenging, if for no other reason than the pigs' insatiable appetites and astoundingly rapid growth. Ordinarily this would also imply a significant expense, but Carpenter and Jacobs possess an unbreakable commitment to resourcefulness.

As the pigs grew hungrier, the humans found more and more sources of free food. The alleys in the nearby Chinatown district of Oakland turned out to be garbage gold mines, teeming with potential nourishment for the livestock. Carpenter and Jacobs would embark on midnight expeditions into putrid trash bins, digging out everything from seafood to layer cake—a particular favorite of the pigs—and driving the loot back to the farm, in a diesel car running on secondhand grease from local restaurants.

After the pig experiment was complete and Carpenter had a kitchen full of charcuterie she'd made herself, she moved on to goats—slightly less cumbersome animals that brought a whole new set of lessons to the eager farmer. She requested the tutelage of

LAND, CA

another culinary expert to learn how to make rennet from her young goats' stomachs, which was then used to make cheese.

As Carpenter mastered the many skills of a full-time homesteader, Ghost Town Farm organically became a gathering place for locals interested in taking steps toward self-sufficiency. She began offering workshops in her garden, running an occasional farm stand where she could sell her produce, and hosting pizza parties, serving homemade pies topped with the garden's seasonal output and fired in a cob oven she'd built on the edge of the lot.

In a region of the country where DIY is a point of cultural pride and almost everyone under forty practices some kind of handicraft, from jam-making to welding, Carpenter is an icon among do-it-yourselfers. Though the Bay Area ranks high among U.S. metropolises for cost of living, Carpenter manages to make it on almost nothing, spending very little money and acquiring the necessities of life from the scads of materials that get discarded throughout the city long before they've lost their utility.

Even the land on which she grows most of her food was, for many years, a free resource that she was able to make use of because nobody else wanted it. Unable to develop the property, the owner permitted her to garden there for free. But when the neighborhood began to change and he thought he might at last to get some money out of the lot, he threatened to uproot the farm. Carpenter endured several seasons of uncertainty, waiting for the day the landlord would lock her out of her garden, but the economic downturn and very slow improvement of neighborhood safety kept development at bay. Finally, in 2010, Carpenter and Jacobs managed to buy the lot themselves, securing the future of Ghost Town Farm.

To celebrate becoming a landowner and closing on a dream she had believed might never be realized, Carpenter did what any true farmer would do: she planted asparagus, a famously long-lived perennial that she was not willing to cultivate when she thought her tenancy on the land might be temporary. In the longer term, she hopes to replace her chain-link fence with espaliered fruit trees around the edge of the property, and someday, perhaps, build herself and Jacobs a freestanding house of their own, at the center of the farm they cultivated from scratch in Ghost Town.

Getting Your Goat

Not every city allows its residents to keep livestock in their yards—in fact, most don't. Backyard chickens and bees are increasingly common, but goats are relatively rare. If you do live in a city that permits a goat or two, or if you live outside the city where zoning is a bit more lenient, testing the waters of goat ownership can be both rewarding and entertaining, though it definitely takes work.

Since space is limited in urban environments, miniature goats are a good bet. La Manchas, Oberhaslis, and Nigerian Dwarfs and African Pygmies are common choices. While the goats do not grow to be very large, it's critical to have enough space for them—ideally no less than sixty-five square feet. Goats can serve a variety of functions on an urban farm. They are excellent composters and generators of fertilizer, as well as satisfactory weed eliminators—though be warned they are not as discriminating as one might hope, so weeds won't be the only thing they eat if presented the right opportunity. As providers of dairy, they also excel, but the goat owner must be prepared to deal with breeding and babies, as well as to learn the skill of deftly milking them every day, and, of course, the culinary art of making goat cheese, ice cream, and other rich dairy products.

WATTLES COMMUNITY GARDEN

LOS ANGELES, CALIFORNIA

Up the slope from Hollywood Boulevard, along the edge of a wide hiking path where Los Angeles residents get their daily exercise, a dense cluster of avocado trees casts a cool shadow in the hot city. Planted in the early twentieth century by a wealthy Midwestern banker named Gurdon Wattles, who came to Southern California to build himself a winter getaway, the trees predate much of the construction on the nearby streets.

Wattles himself died in 1932, but his mansion and estate live on, now as a registered cultural monument and the property of the Los Angeles Recreation and Parks Commission. And on the 4.2 acres of land surrounding the mansion, the well-established avocado orchard represents just a portion of the edible profusion now cultivated by hundreds of neighborhood gardeners.

Wattles Farm was founded in 1975, making it the oldest community garden in Los Angeles. Entering the front gates feels like stepping into an enchanted, semitropical forest, where banana and papaya trees share soil with common cooking herbs and tomatoes. The perimeter fence is lined with figs, citrus, stone fruit and the occasional guava. Many of the gardeners at Wattles are Eastern European immigrants who were accustomed to growing their own food in their home countries, and who embraced a chance to cultivate a small patch of beets, dill, and other staple ingredients of their native cuisine. In the Southern California climate, the fifteen-foot-square plots teem with vegetables all year round—in many cases more than enough to feed a single person or even a small family.

Becoming a community gardener at Wattles Farm is no casual affair. The waiting list is long, and potential members must be serious about tending their plots and respecting the common territory in between. Regular volunteer workdays ensure that the pathways that crisscross the property stay well maintained and navigable, even during wet weather. While gardeners roam almost all of the acreage, the avocado orchard possesses an aura of secrecy—visitors are told to keep out, and members abide by strict rules around taking home the coveted fruit.

Though the avocados have been known to produce some contention, members are generally eager to share what they grow. Strike up a conversation with anyone on a sunny afternoon and you're likely to be offered a taste of whatever's ripe, and introduced to fruit and vegetable varietals that wouldn't

grow in many other parts of the country. A few times a year, the master gardeners who manage the farm organize a community dinner for all Wattles members—a chance for neighbors to meet one another, and proud gardeners to show off their harvest.

Wattles Farm is one of more than seventy community gardens in Los Angeles supported by the Los Angeles Community Garden Council, a nonprofit organization that helps neighbors establish and operate gardens. Most of the community plots are geared toward self-reliance and garden education, with the food feeding the families who grow it, as opposed to being sold for profit to markets or CSAs. In a spread-out city like Los Angeles, gardens are an invaluable way to insert healthy, fresh food into regions that have minimal access to supermarkets or farm stands.

In Hollywood, where Wattles is situated, access to good food is not a problem for most people, but many residents still lack space to grow vegetables at home. The large, lush garden tucked into the base of the hills is a precious part of life for the locals who find their way in, and a welcome sight for asphalt-weary hikers passing by on their way to catch some fresh air.

MILAGRO ALLEGRO COMMUNITY GARDEN

LOS ANGELES, CALIFORNIA

One of the most recognizable landmarks in the Highland Park neighborhood of northeast Los Angeles is the massive Highland Theatre sign, which sits atop a vintage cinema building that has operated since the 1920s. But behind the old theater, another important neighborhood hub has sprouted. Milagro Allegro Community Garden could be viewed as the antithesis of a movie house—a place to get out of one's seat and turn up the physical activity by digging into the soil. Less than five years after it was founded, the garden is already a buzzing community gathering place, where novices and pros alike come to grow food for themselves and their families.

Milagro Allegro was established by a small group of neighborhood residents who reflect the racial and ethnic diversity of Highland Park. The project name is a combination of Spanish and Italian words that translate as "happy miracle"—the founders' feeling about turning ten-thousand square feet of long-neglected city land into a food source for the community.

The garden is made up of thirty 6-foot-by-12-foot plots. Milli Macen-Moore, the master gardener in residence, teaches new members how to plant, cultivate, and harvest crops. "Most people who come here are basic gardeners," she says, "Many of them arrive not knowing where food comes from. They come here to find out." She begins with things that thrive in Southern California, like tomatoes, eggplant, peppers, squash, and melons. "We have a lot of cultural diversity in the community," she adds, "So we also grow different types of herbs, many from Mexico and South America, such as epazote, papalo, and oregano."

Many of the members find that their plots quickly deliver a profusion of fresh food—sometimes more than they can eat themselves. But none sell their yield for profit. Instead, the garden hosts weekly produce swaps where neighbors can trade their surplus items and fill ingredient voids. To help gardeners figure out how best to use what they grow, Macen-Moore runs a regular program called Fresh from the Garden, which includes cooking demonstrations and nutritional overviews of seasonal vegetables and fruits, and culminates in a potluck. "People start noticing that they don't have to go to the grocery store as much as they used to," she remarks, "because their neighbors are growing what they need now."

Macen-Moore graduated from the University of California Cooperative Extension's master gardening program in 2009 and quickly set down her roots at Milagro Allegro. The garden has become a destination for hands-on education in growing fresh food. While the primary members live within a relatively close radius, workshop participants come from as far away as Camarillo and San Bernardino County to learn about composting, worm casting, and organic fertilizing. Those who do not have a plot but wish to remain involved often show up for the weekly community work hours on Thursday evenings, for which they are rewarded with an armload of produce from the educational plots.

Milagro Allegro is part of a network of twenty farms around Los Angeles County that comprise the Grow LA Victory Garden Initiative, a certification program that teaches Angelenos how to start their own gardens, then encourages them to establish new community food operations and edible schoolyards. The work in youth education is especially important to the mission of Milagro Allegro, and dear to Macen-Moore's heart. In 2011, the garden partnered with researchers from the University of Southern California and the University of California, Los Angeles, to run a pilot program studying the relationship between childhood obesity and gardening. Over the course of twelve weeks, fourth- and fifth-grade Latino students participated in food-growing classes at Milagro Allegro, at the end of which researchers determined that garden education had a positive impact on weight management in kids.

Due to the popularity of Milagro Allegro in Highland Park, members cycle through in two-year periods, absorbing the abundant knowledge offered by Macen-Moore and her colleagues, then taking their skills on to their own yards or other community projects around Los Angeles. "Many people have gone on to get jobs, started their own Fresh from the Garden programs, and many come back on Thursday nights to help maintain the educational plots," Macen-Moore reports, "If a community gardener comes by one day asking about nematodes, I give a quick lesson on the spot. It's a continuous learning process."

Kathleen

BIOGRAPHIES

ALLISON ARIEFF writes about design, architecture, and sustainability for the *New York Times* and the *Atlantic Cities*, and is editor of the *Urbanist*, the magazine of SPUR (San Francisco Planning and Urban Research Association). She has a five-hundred-square-foot urban "farm" in her backyard in San Francisco.

MAKALÉ FABER CULLEN is a cultural anthropologist. She has documented and presented traditional artisans for the Smithsonian's Center for Folklife and Cultural Heritage, run educational programs for City Lore, and documented North American agricultural biodiversity while program director of Slow Food USA. She has worked closely on supply-chain enhancement for both institutional food service and retail grocers and designed the Urban Solutions for the Environment (USE) "Urban Food System: Greening Certificate" program for the Center for Economic and Workforce Development at Kingsborough Community College. She is the proprietor of the shop Wilderness of Wish and heads an ethnographic research consultancy collaborative, lore. She and her husband garden at Warren-St.Marks Community Garden in Brooklyn.

RUPAL SANGHVI is a public health expert and founder of HealthxDesign, which explores the role of design to optimize health outcomes. Rupal is a fellow at the Design Trust for Public Space, where she is working on the Five Borough Farm project to propose a New York City–wide plan for urban agriculture, including a metrics framework for demonstrating the multiple and interrelated benefits of urban agriculture activity. She is curious and optimistic about current opportunities to tap the latent potential of linking food, culture, urban agriculture, communities, and health.

NICOLA TWILLEY is author of the blog *Edible Geography* and director of Studio-X NYC, part of the Columbia University Graduate School of Architecture, Planning, and Preservation's global network of advanced research laboratories for exploring the future of cities. She is curator of a forthcoming exhibition (spring 2012) at the Center for Land Use Interpretation that explores North America's spaces of artificial refrigeration; cofounder of the Foodprint Project; and former food editor at *GOOD* magazine.

ALISSA WALKER writes about design, architecture, cities, transportation, and walking for many publications, including *GOOD*, *Fast Company*, and *Dwell*, and is the associate producer for the KCRW public radio show "DnA: Design and Architecture." She lives in a royal-blue house in the Silver Lake neighborhood of Los Angeles, where she throws ice cream socials, tends to a drought-tolerant garden, rides a Creamsicle-colored public bike, writes infrequently on her blog, *Gelatobaby*, and relishes life in Los Angeles without a car.

ACKNOWLEDGMENTS

Like the community-supported farms featured in these pages, this book was driven forward by a small but mighty community of colleagues and friends. For their unyielding encouragement and thoughtful reading, I'm grateful to my parents and family, Carissa Bluestone, and my editor, Andrea Danese. My admiration and appreciation also go to the five brilliant essayists in this book: Allison Arieff, Makalé Faber Cullen, Rupal Sanghvi, Nicola Twilley, and Alissa Walker. Thanks to Matthew Benson for several adventure-filled trips into the field, and all the farmers we found there; Jeff Zerger and Emily Leslie for their agro-ecological expertise; Sarah Gifford at Abrams for her design; and Alexis C. Madrigal, my collaborator for life.

HOLLYGROVE MARKET AND FARM (PAGE 170)

LOCATION

Editor: Andrea Danese
Designer: Sarah Gifford
Production Manager: Jules Thomson

Library of Congress
Cataloging-in-Publication Data:

Rich, Sarah.
Urban farms / by Sarah Rich ; photography by Matthew Benson.
p. cm.
ISBN 978-1-4197-0199-3
1. Urban agriculture—United States. 2. Farms—United States—Case studies. I. Title.
S494.5.U72U53 2012
635.9'77—dc23

2011045759

Printed and bound in Hong Kong, China
10 9 8 7 6 5 4 3 2 1

115 West 18th Street
New York, NY 10011
www.abramsbooks.com